I-Hsien Ting, Hui-Ju Wu, and Tien-Hwa Ho (Eds.)

Mining and Analyzing Social Networks

Studies in Computational Intelligence, Volume 288

Editor-in-Chief
Prof. Janusz Kacprzyk
Systems Research Institute
Polish Academy of Sciences
ul. Newelska 6
01-447 Warsaw
Poland
E-mail: kacprzyk@ibspan.waw.pl

Further volumes of this series can be found on our homepage: springer.com

Vol. 267. Ivan Zelinka, Sergej Celikovský, Hendrik Richter, and Guanrong Chen (Eds.)
Evolutionary Algorithms and Chaotic Systems, 2009
ISBN 978-3-642-10706-1

Vol. 268. Johann M.Ph. Schumann and Yan Liu (Eds.)
Applications of Neural Networks in High Assurance Systems, 2009
ISBN 978-3-642-10689-7

Vol. 269. Francisco Fernández de de Vega and Erick Cantú-Paz (Eds.)
Parallel and Distributed Computational Intelligence, 2009
ISBN 978-3-642-10674-3

Vol. 270. Zong Woo Geem
Recent Advances In Harmony Search Algorithm, 2009
ISBN 978-3-642-04316-1

Vol. 271. Janusz Kacprzyk, Frederick E. Petry, and Adnan Yazici (Eds.)
Uncertainty Approaches for Spatial Data Modeling and Processing, 2009
ISBN 978-3-642-10662-0

Vol. 272. Carlos A. Coello Coello, Clarisse Dhaenens, and Laetitia Jourdan (Eds.)
Advances in Multi-Objective Nature Inspired Computing, 2009
ISBN 978-3-642-11217-1

Vol. 273. Fatos Xhafa, Santi Caballé, Ajith Abraham, Thanasis Daradoumis, and Angel Alejandro Juan Perez (Eds.)
Computational Intelligence for Technology Enhanced Learning, 2010
ISBN 978-3-642-11223-2

Vol. 274. Zbigniew W. Raś and Alicja Wieczorkowska (Eds.)
Advances in Music Information Retrieval, 2010
ISBN 978-3-642-11673-5

Vol. 275. Dilip Kumar Pratihar and Lakhmi C. Jain (Eds.)
Intelligent Autonomous Systems, 2010
ISBN 978-3-642-11675-9

Vol. 276. Jacek Mańdziuk
Knowledge-Free and Learning-Based Methods in Intelligent Game Playing, 2010
ISBN 978-3-642-11677-3

Vol. 277. Filippo Spagnolo and Benedetto Di Paola (Eds.)
European and Chinese Cognitive Styles and their Impact on Teaching Mathematics, 2010
ISBN 978-3-642-11679-7

Vol. 278. Radomir S. Stankovic and Jaakko Astola
From Boolean Logic to Switching Circuits and Automata, 2010
ISBN 978-3-642-11681-0

Vol. 279. Manolis Wallace, Ioannis E. Anagnostopoulos, Phivos Mylonas, and Maria Bielikova (Eds.)
Semantics in Adaptive and Personalized Services, 2010
ISBN 978-3-642-11683-4

Vol. 280. Chang Wen Chen, Zhu Li, and Shiguo Lian (Eds.)
Intelligent Multimedia Communication: Techniques and Applications, 2010
ISBN 978-3-642-11685-8

Vol. 281. Robert Babuska and Frans C.A. Groen (Eds.)
Interactive Collaborative Information Systems, 2010
ISBN 978-3-642-11687-2

Vol. 282. Husrev Taha Sencar, Sergio Velastin, Nikolaos Nikolaidis, and Shiguo Lian (Eds.)
Intelligent Multimedia Analysis for Security Applications, 2010
ISBN 978-3-642-11754-1

Vol. 283. Ngoc Thanh Nguyen, Radoslaw Katarzyniak, and Shyi-Ming Chen (Eds.)
Advances in Intelligent Information and Database Systems, 2010
ISBN 978-3-642-12089-3

Vol. 284. Juan R. González, David Alejandro Pelta, Carlos Cruz, Germán Terrazas, and Natalio Krasnogor (Eds.)
Nature Inspired Cooperative Strategies for Optimization (NICSO 2010), 2010
ISBN 978-3-642-12537-9

Vol. 285. Roberto Cipolla, Sebastiano Battiato, and Giovanni Maria Farinella (Eds.)
Computer Vision, 2010
ISBN 978-3-642-12847-9

Vol. 286. Alexander Bolshoy, Zeev (Vladimir) Volkovich, Valery Kirzhner, and Zeev Barzily
Genome Clustering, 2010
ISBN 978-3-642-12951-3

Vol. 287. Dan Schonfeld, Caifeng Shan, Dacheng Tao, and Liang Wang (Eds.)
Video Search and Mining, 2010
ISBN 978-3-642-12899-8

Vol. 288. I-Hsien Ting, Hui-Ju Wu, Tien-Hwa Ho (Eds.)
Mining and Analyzing Social Networks, 2010
ISBN 978-3-642-13421-0

I-Hsien Ting, Hui-Ju Wu, Tien-Hwa Ho (Eds.)

Mining and Analyzing Social Networks

Springer

Dr. I-Hsien Ting
Department of Information
Management, No. 700
National University of Kaohsiung
Kaohsiung University Rd.
Kaohsiung, 811
Taiwan 5.
E-mail: iting@nuk.edu.tw

Dr. Tien-Hwa Ho
Department of Information
Management, No. 700
National University of Kaohsiung
Kaohsiung University Rd.
Kaohsiung, 811
Taiwan 5

Dr. Hui-Ju Wu
Department of Information
Management, No. 700
National University of Kaohsiung
Kaohsiung University Rd.
Kaohsiung, 811
Taiwan 5

ISBN 978-3-642-13421-0 e-ISBN 978-3-642-13422-7

DOI 10.1007/978-3-642-13422-7

Studies in Computational Intelligence ISSN 1860-949X

Library of Congress Control Number: 2010928121

Typeset & Cover Design: Scientific Publishing Services Pvt. Ltd., Chennai, India.

Printed on acid-free paper

9 8 7 6 5 4 3 2 1

springer.com

Preface

Mining social networks has now becoming a very popular research area not only for data mining and web mining but also social network analysis. Data mining is a technique that has the ability to process and analyze large amount of data and by this to discover valuable information from the data. In recent year, due to the growth of social communications and social networking websites, data mining becomes a very important and powerful technique to process and analyze such large amount of data. Thus, this book will focus upon Mining and Analyzing social network.

Some chapters in this book are extended from the papers that presented in MSNDS2009 (the First International Workshop on Mining Social Networks for Decision Support) and SNMABA2009 ((The International Workshop on Social Networks Mining and Analysis for Business Applications)). In addition, we also sent invitations to researchers that are famous in this research area to contribute for this book. The chapters of this book are introduced as follows:

In chapter 1-Graph Model for Pattern Recognition in Text, Qin Wu et al. present a novel approach that uses a weighted directed multigraph for text pattern recognition. In the proposed methodology, a weighted directed multigraph model has been set up by using the distances between the keywords as the weights of arcs as well a keyword-frequency distance based algorithm has also been introduced. Case studies are also included in this chapter to show the performance is better than traditional means.

In chapter 2-Information Retrieval in Wikis using an Ontology, Carlos Miguel Tobar et al. presented an system which is designed based on the ideas from the semantic Web combined with adaptive mechanisms and a modification of the classic vector model for information retrieval. This system can be used to extract relevant information from huge amount of txt, such as wiki.

In chapter 3-Ego-centric Network Sampling in Viral Marketing Applications, Huaiyu (Harry) Ma et al. describe a study about ego-centric network sampling to show the network structure can be captured accurately. The Stanford-Berkeley network to show that the approach can capture the underlying structure with a minimal amount of data.

In chapter 4- Integrating SNA and DM Technology into HR Practice and Research: Layoff Prediction Model, Hui-Ju Wu et al. proposed a new application direction to combine the techniques of SNA and DM into the research area of Human Resource Management. In this chapter, a valuable dataset has been used to analyze the social structure in a organization and by this to discover the reasons behind layoff.

In chapter 5-Actor Identification in Implicit Relational Data Sources, Michael Farrugia and Aaron Quigley presents a study of a range of techniques that can be employed to identify unique actors when inferring networks from non explicit network data sets. They also present methods for unique node identification of social network actors in a business scenario. A real world case study has also been included in this chapter.

In chapter 6- Perception of Online Social Networks, Travis Green and Aaron Quigley examine data derived from an application on Facebook.com that investigates the relations among members of their online social network. It confirms that online social networks are more often used to maintain weak connections but that a subset of users focus on strong connections, determines that connection intensity to both connected people predicts perceptual accuracy, and shows that intra-group connections are perceived more accurately.

In chapter 7- Ranking Learning Entities on the Web by Integrating Network-based Features, Yingzi Jin et al. propose an algorithm to generate and integrate network-based features systematically from a given social network that is mined from the world-wide web. After learning a model for explaining target rankings researchers' productivity based on social networks confirms the effectiveness of our models. This chapter specifically examines the application of a social network that exemplifies the advanced use of social networks mined from the web.

In chapter 8-Discovering Proximal Social Intelligence for uality Decision Support, Yuan-Chu Hwang focus on discovering the proximal social intelligence or quality decision support. The author illustrates a case of leisure recommendatory e-service for bicycle exercise entertainment in Taiwan as well as introduces the proximity e-service as well as its theoretical support.

In chapter 9- Discovering User Interests by Document Classification, Loc Nguyen propose I propose a new approach for discovering user interest based on document classification. The basic idea is to consider user interests as classes of documents. The process of classifying documents is also the process of discovering user interests.

In chapter 10- Network Analysis of Opto-electronics Industry Cluster: A Case of TAIWAN, Ting-Lin LEE provides a study to describe supply chain relationships networks of opto-electronics industry in STSP as fully as possible, tease out the prominent patterns in such networks, and discover what effects these relationships and networks have on organizations performance. The results of this study contribute to a better understanding of how firms can utilize network benefits to enhance their innovation performance. Furthermore, "coreness centrality" is the most interpretable position variable for innovation performance.

In summary, this book's content sets out to highlight the trends in the research area in Mining and Analysis of Social Networks. Through integrating the two research areas of social networks analysis and data mining, more and more applications and research ideas can be rised.

I-Hsien Ting
Hui-Ju Wu
Tien-Hwa Ho

Contents

Graph Model for Pattern Recognition in Text

Qin Wu, Eddie Fuller, and Cun-Quan Zhang

Abstract. In this paper, we propose a novel approach that uses a weighted directed multigraph for text pattern recognition. Instead of the traditional model which is based on the frequency of keywords for text classification, we set up a weighted directed multigraph model using the distances between the keywords as the weights of arcs. We then developed a keyword-frequency-distance-based algorithm which not only utilizes the frequency information of keywords but also their ordering information. We applied this new idea to the detection of plagiarized papers and the detection of fraudulent emails written by the same person. The results on these case studies show that this new method performs much better than traditional methods.

1 Introduction

For text archives containing a large number of documents, determining the similarity of documents is an area of research that has seen a great deal of activity in recent years. With the advent and ubiquity of internet communication the search for related documents plays an important role in such applications as search, detection of fraud, and the detection of conspiring groups. Term frequency has long been used as a tool for estimating the probabilistic distribution of features in a document. A number of applications have been developed including language modeling [15], feature selection [25, 19], and term weighting [8, 16]. Based on the term frequency information, documents can be classified by several clustering methods such as decision trees

Qin Wu · Eddie Fuller · Cun-Quan Zhang
Department of Mathematics, West Virginia University,
Morgantown, WV 26506-6310, USA

I.-H. Ting et al. (Eds.): Mining and Analyzing Social Networks, SCI 288, pp. 1–20.
springerlink.com

[1], neural networks [18, 13], Bayesian methods [12, 24], or support vector machines [21, 7, 22].

The term frequency method is an effective approach if a rough classification of documents based on their subjects or themes. However, if one would like to further determine the similarity of writing patterns or determine the authorship of documents, the traditional term frequency method will provide only very rough estimates with little accuracy or reliability. The main drawback of the term frequency method is the fact that it relies on a bag-of-words [6, 10, 20] approach. It implies feature independence, and disregards any dependencies that may exist between words in the text. The bag-of-words model may not be the best technique to capture keyword importance. If the text structure information could be preserved properly at the same time, it would lead to a better keyword weighting scheme [5].

In this paper, we introduce a new approach that exploits not only the keyword frequency but also their location and ordering. We represent a document as a weighted directed multigraph by taking keywords as the vertices and constructing arcs whose weighting contains the relation information of a keyword to other keywords. The adjacency matrix of the graph induces a signature vector for the document. A clustering method is then applied to the set of signature vectors for grouping similar documents into clusters. With this new approach, we are able to evaluate the similarity between any two documents from a set of text documents within the SAME category.

A set of detailed algorithms for the estimation of signature vectors and clustering are presented in this paper. This algorithm has been applied to two sets of sample documents.

1. Nigerian Fraud Emails, each of which has the same topic: to transfer money into some bank accounts in order to receive lager sum of payback.
2. Papers in academic journals in graph theory, some of which are known to be plagiarized.

Each group is in one category, and therefore, keywords may appear with similar frequencies. The traditional method of sorting documents by keyword-frequency is able to filter this group out off a lager subset of documents with many different subjects. However, by considering the ordering and location of keywords, we are able to further evaluate their similarity within their own group, i.e. to classify fraudulent emails authored by the same person or copy-pasted types with slight modification, or to identify the plagiarized papers.

In next section, we describe the schema for representing a document as a weighted directed multigraph. Section 3 discusses the computation complexity. In section 4, we present some application examples of our algorithm. Finally, in section 5, the conclusion is presented and future research problems are outlined.

2 Graph Model for Pattern Recognition

The overall approach of this algorithm begins with the identification of a set of relevant keywords. Once these are selected, we then aggregate the relative distances of the keywords with a document. This in turn is used to construct a weighted directed multigraph that generates representing vectors for each document in a high dimensional feature space. These vectors can then be used to determine similarity values for any pair of documents.

2.1 Summary of Our Method

Step 1: Using a weighted directed multigraph to find a signature vector for each document.
Step 2: Calculate the similarities between any two documents via their signature vectors.
Step 3: Using Quasi-Clique Merge clustering method to classify all documents.

We will explain the details of each step by a simple example.

2.2 Details of the Step 1

To have a clear view of the algorithm, we will use the example illustrated in Figure 1 [26] to explain the procedure.

From: Mr. Davies Manga
Tel: 234 - 80- 33345356
E-mail: daviesmanga@qrio.com
Dear Friend,

This mail might be a surprise since we have not met, but this can be remedied as we proceed in this transaction.
I am the personal assistance to the Nigerian late Minister of Sports Mr. Ishaya Mark Aku who died on a plane crash which took place on the 4th April 2002. Prior to his death, I assisted him in depositing the sum of US$18.5 million (Eighteen Million Five Hundred Thousand U. S Dollars) in a local bank in my country (Nigeria).
I solicit your assistance in receiving these funds on our behalf into your account for further investment into a profitable venture in your country. The details of the transfer will be forwarded to you once I receive the assurance of your full cooperation.
You will be entitled to 15% of the total sum as soon as this fund is transferred into your account.
The nature of your business does not necessarily matter in this transaction. Awaiting your urgent response through my above telephone and e-mail address

Yours truly,
Mr. Davies Manga.

Fig. 1 A fraudulent email

2.2.1 Record the Keyword Information Appeared in the Document

For a given document, the following steps are applied to it. Suppose we have already chosen a set of words as keywords, say $\mathcal{K} = \{\overline{K_1}, \overline{K_2}, \cdots, \overline{K_m}\}$. Record every keyword and its position in the document. We will use the following notation:

- k_i represents one of the keyword in the keyword set $\mathcal{K}$.
- i represents the order of the keyword appearing in the document. (It is possible that k_i, k_j are the same element in $\mathcal{K}$.)
- $\widetilde{m}$ represents the total number of keywords appearing in the document.
- p_i is a integer, which represents the total number of the words from the beginning of the document to the word k_i.

In addition we record the frequency of each keyword at the same time. Thus we have the Keyword-Position information table(Table 1).

Table 1 Keyword-Position table

Keyword appears in the document	Position in the document
k_1	p_1
k_2	p_2
$\vdots$	$\vdots$
$k_{\widetilde{m}}$	$p_{\widetilde{m}}$

The details of this process are illustrated as follows (with Figure 1 as an example). For this example, we use the keyword set: {*bank, fund, account, transfer*}. Its Keyword-Position information is listed in Table 2. Frequency information of each keyword for the given example (Figure 1) is listed in Table 3.

Table 2 Keyword-Position information of the email inFigure 1

Keyword appears in the document	Position in the document
bank	91
fund	103
account	109
transfer	124
fund	153
transfer	155
account	158

Table 3 Keyword-Frequency information of the email in Figure 1

keyword	frequency
bank	1
fund	2
account	2
transfer	2

2.2.2 Construct a Weighted Directed Multigraph

For a given document D and a set of keywords $\mathcal{K}$, let G_m be a weighted directed multigraph G_m with the vertex set $\mathcal{K} = \{\overline{K_1}, \overline{K_2}, \cdots, \overline{K_m}\}$ constructed as follows.

Suppose that $k_1, \cdots, k_s$ is the sequence of words such that

(1) each k_μ is a keyword of the given set $\mathcal{K}$,
(2) $k_1, \cdots, k_s$ appear in the document D in this order,
(3) the position of the word k_μ in the document D is p_μ (the p_μ-th word in the document D, $(1 \leq p_1 < \cdots < p_s)$.

Add an arc from the vertex k_i to the vertex k_j with the weight $w\_m_{ij} = p_j - p_i + 1$, which is the distance from the word k_i to the word k_j in the document D.

Note that if k_i and k_j are the same element of the set $\mathcal{K}$, they are the same vertex in the graph.

A large weight for a given arc indicates that the corresponding pair of keywords are relatively far away from each other and, therefore, their logical connection are relatively "weak" in the document. Thus, we may ignore those arcs with large weights. (We choose a *threshold* = 200 in our example in Figure 1 and delete any arc with weight greater than 200.)

Note that the resulted weighted directed multigraph may contain not only parallel arcs but also loops.

For the given example (Figure 1), its corresponding weighted directed multigraph is Figure 2.

2.2.3 Simplification of Representing Graphs

The weighted directed multigraph G_m constructed in the previous step is further simplified as follows (a directed graph G_s is constructed from G_m, in which, parallel arcs are combined).

Let $E_{ij} = \{k_\mu k_\nu | \ k_\mu = \overline{K_i} \ \& \ k_\nu = \overline{K_j}\}$, which is the set of all arcs from the vertex $\overline{K_i}$ to the vertex $\overline{K_j}$ of the weighted directed multigraph G_m.

Let $\mathcal{K} = \{\overline{K_1}, \overline{K_2}, \cdots, \overline{K_m}\}$ be the vertex set of the new directed graph G_s. For each pair of vertices $\overline{K_i}$ and $\overline{K_j}$ ($i, j = 1, 2, ..., m$), if $E_{ij} \neq \emptyset$, put an arc e_{ij} from $\overline{K_i}$ to $\overline{K_j}$. The weight of the arc $e_{ij} = \overline{K_i K_j}$ is calculated as follows,

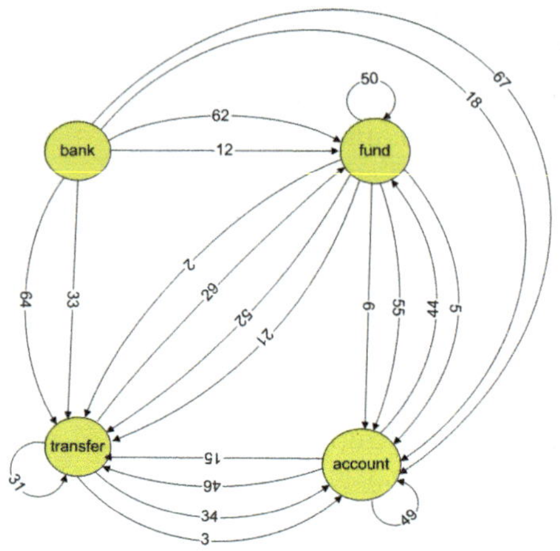

Fig. 2 The weighted, directed multigraph of the email in Figure 1

$$w\_s_{ij} = \sum_{k_\mu k_\nu \in E_{ij}} \frac{1}{w\_m_{\mu\nu}}, \qquad \text{if } E_{ij} \neq \emptyset.$$

The terms $\frac{1}{w\_m_{\mu\nu}}$ are constructed so that when two terms are closer to each other the reciprocal of their small relative distance will contribute more strongly to the summation. When terms are farther apart, the reciprocal will be small and so these terms will contribute less.

The simplified directed graph G_s of the given example is illustrated in Figure 3.

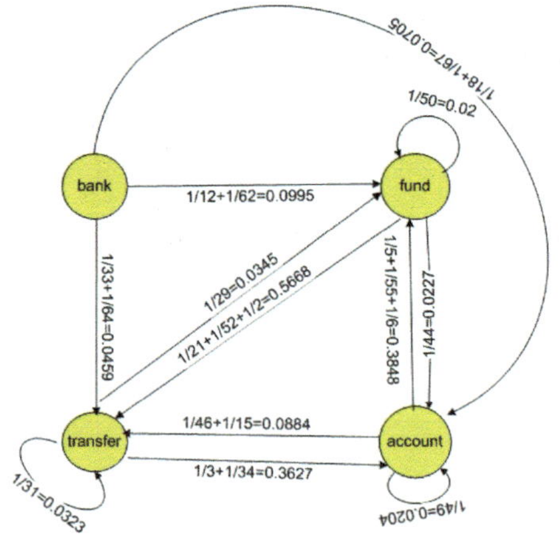

Fig. 3 The simplified directed graph of the email in Figure 1

2.2.4 Create a Signature Vector to Represent the Input Email

Now we create a signature vector to represent an input email by the frequency information of the keywords and the simplified weighted graph information.

1. We use f_i to denote the frequency of the keyword $\overline{K_i}$ in the document. Use $F(D) = [f_1, f_2, ..., f_m]$ denote the frequency vector of the document D.
2. We use the adjacency matrix to represent the simplified weighted directed graph G_s.

Let $w\_s_{ij} = 0$ if there is no arc from the vertices $\overline{K_i}$ to $\overline{K_j}$.

$$W(D) = \begin{bmatrix} w\_s_{11} & w\_s_{12} & \cdots & w\_s_{1m} \\ w\_s_{21} & w\_s_{22} & \cdots & w\_s_{2m} \\ \vdots & \vdots & \ddots & \vdots \\ w\_s_{m1} & w\_s_{m2} & \cdots & w\_s_{mm} \end{bmatrix}.$$

Then we rewrite it as an $(m \times m)$ vector.

$$\widetilde{W}(D) = [w\_s_{11}, w\_s_{12}, ..., w\_s_{1m}, w\_s_{21}, w\_s_{22}, ...w\_s_{2m}, ..., w\_s_{mm}].$$

Let $R(D) = [F(D), \widetilde{W}(D)]$. The vector $R(D)$ contains not only the frequency information of the keywords, but also the structure information of the document. It is used as the signature vector of the document.

Again, corresponding to the given example (Figure 1), we have

$$F(D) = [1, 2, 2, 2]$$

$$W(D) = \begin{bmatrix} 0 & 0.0995 & 0.0705 & 0.0459 \\ 0 & 0.0200 & 0.3848 & 0.5668 \\ 0 & 0.0227 & 0.0204 & 0.0884 \\ 0 & 0.0345 & 0.3627 & 0.0323 \end{bmatrix}$$

$$R(D) = [1,\ 2,\ 2,\ 2,\ 0,\ 0.0995,\ 0.0705,\ 0.0459,\ 0,\ 0.0200,\ 0.3848,\ 0.5668, \\ 0,\ 0.0227,\ 0.0204,\ 0.0884,\ 0,\ 0.0345,\ 0.3627,\ 0.0323].$$

2.3 *Details of the Step 2*

2.3.1 Find Signature Vectors for All Documents

Repeat the process of the Step 1, we create signature vectors for all documents.

Let $R(D_i)$ be the signature vector of the i-th document.

Let $M = [R(D_1), R(D_2), ..., R(D_{n-1}), R(D_n)]^T$, then M is an $n \times (m + m^2)$ matrix (n is the total number of the documents, m is the cardinality of the keywords set $\mathcal{K}$. Each row of the matrix represents a document.

2.3.2 Normalization of the Matrix

We normalize the matrix M with respect to the columns for the purpose of the compatibility in every dimension. We denote the normalized matrix as $\widetilde{M} = [\widetilde{R}(D_1), \widetilde{R}(D_2), ..., \widetilde{R}(D_{n-1}), \widetilde{R}(D_n)]^T$. And the details of the normalization is presented in next section.

2.3.3 Similarity

The similarity S_{ab} between any two documents $\overline{D}_a, \overline{D}_b$ is determined by the cosine similarity as follows

$$S_{ab} = \frac{|\widetilde{R}(D_a) \cdot \widetilde{R}(D_b)|}{|\widetilde{R}(D_a)| \cdot |\widetilde{R}(D_b)|}$$

where $\widetilde{R}(D_a)$, $\widetilde{R}(D_b)$ are the normalized signature vectors of the documents $\overline{D}_a, \overline{D}_b$.

2.4 Details of the Step 3

A variety of different clustering algorithms have been developed and implemented in popular statistical software packages. A general review of cluster analysis can be found in many references, for instance, [4, 3, 11], etc. None of these algorithms can, in general, rigorously guarantee to produce a globally optimal clustering for non-trivial objective functions [23].

After calculating the pairwise similarities of all documents, we then classify these documents into different groups by applying the Quasi-Clique Merge(QCM) method to cluster the documents. It is observed that one of the most significant differences between the QCM method and other clustering algorithms is that the QCM method constructs a much smaller hierarchical tree. This tree structure leads to better identification of meaningful clusters since there are fewer subdivisions of the data set due to the impact of irrelevant or improperly interpreted information. Additionally, the QCM method results in multi-membership clustering [14], which preserves some amount of the ambiguity inherent in the data set rather than errantly suppressing it as many other clustering algorithms do.

3 The Algorithm and Complexity Analysis

3.1 Graph Theory Notation and Terminology

Let Σ be the set of the alphabets appearing in the key words which includes the special symbol "$\Box$" as the space character.

Let $\mathcal{D} = \{\overline{D_1}, \overline{D_2}, \cdots, \overline{D_n}\}$ be a set of text documents for pattern detection. Each document $\overline{D_i}$ is a sequence $d_{i,0} \cdots d_{i,t_i}$ consisting of alphabets from the set Σ, where the first and the last character $d_{i,0} = d_{i,t_i} = \Box$, $t_i + 1$ is the length of the document $\overline{D_i}$.

Let $\mathcal{K} = \{\overline{K_1}, \overline{K_2}, \cdots, \overline{K_m}\}$ be the set of selected keywords. For each keyword $\overline{K_i} = k_{i,0} \cdots k_{i,s_i}$, the first and the last character $k_{i,0} = k_{i,s_i} = \Box$, $s_i + 1$ is the length of the keyword $\overline{K_i}$.

Let $G = (V, A)$ be a directed graph with vertex set V and arc set A.

$N^+(v)$ is the set of all out-neighbors of the vertex v. That is, $N^+(v) = \{u \in V(G) : \ vu \in A(G)\}$.

$N^-(v)$ is the set of all in-neighbors of the vertex v. That is, $N^-(v) = \{u \in V(G) : \ uv \in A(G)\}$.

Let $L : A(G) \mapsto \Sigma$ be a labeling of $A(G)$. $L^+(v) = \{l(vu) : \ u \in N^+(v)\}$.

3.2 Construction of Searching Tree

For the purpose of finding keywords efficiently, we use the following algorithm to set up a searching tree for keyword searching.

3.2.1 Algorithm

Input. $\mathcal{K} = \{\overline{K_1}, \overline{K_2}, \cdots, \overline{K_m}\}$: a set of keywords.

Output. A rooted tree (called "searching tree") T: T has a root v_0 and m leaves. Each of the leaf represents a keyword; each arc of T is labeled with a character $\in \Sigma$; for each leaf v_ℓ, let P be the unique directed path from the root v_0 to v_ℓ, the sequence of labels along the path P coincides with characters of the keywords K_ℓ. (Figure 4 gives a simple example of a searching tree.)

Initial step. T has a root v_0 and a vertex v_1, and an arc $v_0 v_1$ with the label $L(v_0 v_1) = \Box$.

$i \leftarrow 1$: i is the keyword index (current keyword $\overline{K_i} = k_{i,0} \cdots k_{i,s_i}$ that is under processing, $k_{i,0} = k_{i,s_i} = \Box$, $s_i + 1$ is the length of the keyword $\overline{K_i}$.)

$\lambda \leftarrow 1$: λ is the level index (the character $k_{i,\lambda}$ is currently under processing, and λ is also current level of the tree that is under construction).

$v \leftarrow v_1$: (the current vertex whose out-neighborhood is under construction.)

Step 1.
Case 1. If $\lambda < s_i$. Consider $N^+(v)$.
Subcase 2-a. If $N^+(v) = \emptyset$, or if $k_{i,\lambda} \notin L^+(v)$, then go to Step 2.
Subcase 2-b. If $k_{i,\lambda} \in L^+(v)$, say, $k_{i,\lambda} = L(vu)$ for some $u \in N^+(v)$, then go to Step 3.

Case 2. If $\lambda = s_i$. (Reach the end of the keyword $\overline{K_i}$.)
If $i < m$ then
$i \leftarrow i + 1$
$\lambda \leftarrow 1$
$v \leftarrow v_1$
and go back to Step 1;
If $i = m$ (reach the last keyword) then go to Step F.

Step 2. (This is the step that adds a directed path with tail at the vertex v).
Add a directed path $u_0 \cdots u_z$ with $\{u_1, \cdots, u_z\}$ as new vertices and $u_0 = v$ and

$$l(u_0u_1) = k_{i,\lambda},\ l(u_1u_2) = k_{i,\lambda+1},\ \cdots,\ l(u_{z-1}u_z) = k_{i,s_i}.$$

Then
$\lambda \leftarrow s_i$
and go to Step 1.

Step 3. (In this step, an existing arc vu will be used since $l(vu) = k_{i,\lambda}$).
$\lambda \leftarrow \lambda + 1$,
$v \leftarrow u$,
go to Step 1.

Step F. Final step: Output.

3.2.2 Complexity

Let $len_{\mathcal{K}} = \sum_{i=1}^{m} |\overline{K_i}|$ denote the total length of all keywords in $\mathcal{K}$. Steps 1 - 3 form a loop that repeats $len_{\mathcal{K}}$ times. For Case 1, and Subcase 2-a, each costs 1 unit for each character of $\overline{K_i}$; for Subcases 2-b, it costs at most $|\Sigma|$ units (for comparisons). For each subcase, an iteration of Step 2 or 3 is followed and afterward, return back to Step 1 for another loop.

Hence, the complexity of constructing the searching tree is $O(len_{\mathcal{K}})$.

Remark: since we only build up this tree once in the whole procedure, the complexity of constructing the searching tree will not be counted into the total complexity.

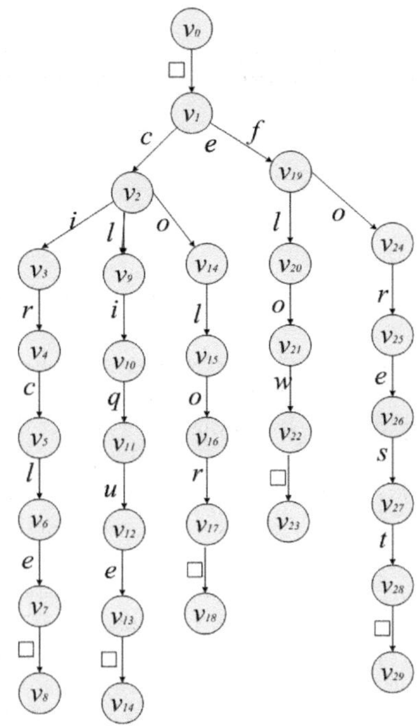

Fig. 4 A searching tree with keyword {circle, clique, color, flow, forest}

3.3 Keyword Searching and Location in Documents

3.3.1 Algorithm

Let $\mathcal{K} = \{\overline{K_1}, \overline{K_2}, \cdots, \overline{K_m}\}$ be the set of selected keywords. A keyword searching tree T was constructed in Algorithm 3.2.1 ready for use. Let $\Theta(T)$ be the set of leaves of the rooted tree T. For the sake of convenience, each leaf of T is denoted by its corresponding keyword. That is, $\Theta(T) = \mathcal{K} = \{\overline{K_1}, \overline{K_2}, \cdots, \overline{K_m}\}$.

Input. A text document $\overline{D_i} = d_{i,0} \cdots d_{i,t_i}$ where the first and the last character $d_{i,0} = d_{i,t_i} = \square$.

Output. The position sets of each keyword in the document. Each keyword $\overline{K_\mu}$ is associated with a set $\mathcal{P}(\overline{K_\mu})$ of integers, where: $p \in \mathcal{P}(\overline{K_\mu})$ if and only if the keyword $\overline{K_\mu}$ appears in the document $\overline{D_i}$ at position p.

Initial Step. $j \leftarrow 0$ (the character $d_{i,j}$ of the document $\overline{D_i}$ is currently in iteration).

$v \leftarrow v_0$

$\mathcal{P}(\overline{K_\mu}) \leftarrow \emptyset$, for each $\overline{K_\mu}$.

$w \leftarrow 1$: w is the position of current word in the document.

Step 1.
Case 1. If $j < t_i$, go to Step 2.
Case 2. If $j = t_i$, go to Step F.

Step 2.
Case 1. If $d_{i,j} \in L^+(v)$, say $l(vu) = d_{i,j}$ where $u \in N^+(v)$, then go to Step 3.
Case 2. If $d_{i,j} \notin L^+(v)$, then go to Step 4.

Step 3.
Case 1. If $N^+(u) \neq \emptyset$ (u is not a leave of the tree T), then
$v \leftarrow u$,
$j \leftarrow j+1$,
go to Step 1.

Case 2. If $N^+(u) = \emptyset$ (u is a leave of the tree T), then
$\mathcal{P}(u) \leftarrow \mathcal{P}(u) \cup \{w\}$,
$v \leftarrow v_0$,
$j \leftarrow j+1$,
$w \leftarrow w+1$,
go to Step 1.

Step 4.
Case 1. If $d_{ij} \neq \square$,
$j \leftarrow j+1$
and go back to Step 4;

Case 2. If $d_{ij} = \square$,
$v \leftarrow v_0$,
$w \leftarrow w+1$,
Go to Step 1.

Step F. Output: $\mathcal{P}(\overline{K_\mu})$, for each $\overline{K_\mu} \in V_L$.

3.3.2 Complexity

Each character $d_{i,j}$ of the document $\overline{D_i}$ is compared with $N^+(v)$ or $N^+(v) \cup N^+(v_0)$ for some vertex $v \in V(T)$. That is, it costs at most $(|\Sigma|+1)$ units for comparisons. So, the total cost is $(|\Sigma|+c) \times t_i$ where c is a small constant cost for re-indexing of j, v and updating the records $\mathcal{P}(\overline{K_\mu})$.

Thus, the complexity of keyword searching is $O(t_i)$, where t_i is the length of the input document D_i.

3.4 Signature Vector of a Document

The *signature vector* $R(\overline{D_i})$ for a given document $\overline{D_i}$ is to be calculated in this section.

Input: The collection of sets $\mathcal{P}(\overline{K_\mu})$ for all keyword $\overline{K_\mu}$ (provided in Algorithm 3.3.1).
Output: An $1 \times (m+m^2)$-vector $R(\overline{D_i})$.

Calculation:

Let $F(\overline{D_i}) = [f_\mu] = [f_1, \cdots, f_m]$ be a $(1 \times m)$-matrix where $f_\mu = |\mathcal{P}(\overline{K_\mu})|$.
Let $W(\overline{D_i}) = [\alpha_{\mu,\nu}]$ be an $(m \times m)$-matrix with

$$\alpha_{\mu,\nu} = \sum \frac{1}{p_\mu - p_\nu + 1}$$

where the summation is over all pairs $p_\mu \in \mathcal{P}(\overline{K_\mu})$ and $p_\nu \in \mathcal{P}(\overline{K_\nu})$ with $p_\mu > p_\nu$.

Note: this is not a symmetric matrix, parallel arcs in opposite directions in the graph are considered differently.

Let $R(\overline{D_i}) = [F(\overline{D_i}), \widetilde{W}(\overline{D_i})]$. Here we rewrite the matrix $W(\overline{D_i})$ as a $1 \times m^2$ row vector $\widetilde{W}(\overline{D_i})$.

Complexity: It costs $|\mathcal{P}(\overline{K_\mu})|$ for the calculation of f_μ, and it costs $|\mathcal{P}(\overline{K_\mu})||\mathcal{P}(\overline{K_\nu})|$ for the calculation of $\alpha_{\mu,\nu}$ and $\alpha_{\nu,\mu}$ for every $\mu, \nu \in \{1, \cdots, m\}$. So, the complexity is $O(m^2\phi^2)$, where ϕ = average of $|\mathcal{P}(\overline{K_\mu})|$ (the average appearance of a keyword in the document $\overline{D_i}$).

3.5 Similarity Calculation

Let $\mathcal{D} = \{\overline{D_1}, \cdots, \overline{D_n}\}$ be a set of documents.

3.5.1 Date Normalization

Input: Let $R(\overline{D_i})$ be the $(1 \times (m+m^2))$ vector calculated in Section 3.4. Additionally let $M(\mathcal{D}) = [\beta_{i,j}]$ be the $(n \times (m+m^2))$-matrix, with the i-th row the signature vector $R(\overline{D_i})$, so that $\beta_{i,j}$ is the j-th component of the signature vector $R(\overline{D_i})$.
Calculation and output: For each $j \in \{1, \cdots, m+m^2\}$, let A_j be the average of all cells in the j-th column of the matrix $M(\mathcal{D})$.

$$\widetilde{M}(\mathcal{D}) := [\widetilde{\beta}_{i,j}] = [\frac{\beta_{i,j}}{A_j}].$$

Complexity: $O(nm^2)$

3.5.2 Similarity

The similarity between two documents $\overline{D_a}$ and $\overline{D_b}$ is then calculated as

$$s_{ab} = \frac{\widetilde{R}(\overline{\mathcal{D}}_a) \cdot \widetilde{R}(\overline{\mathcal{D}}_b)}{|\widetilde{R}(\overline{\mathcal{D}}_a)| \cdot |\widetilde{R}(\overline{\mathcal{D}}_b)|}$$

where $\widetilde{R}(\overline{D_i})$ is the i-th row of the matrix $\widetilde{M}(\mathcal{D})$, namely, the normalized signature vector of the document $\overline{D}_i$.

Complexity: $O(n^2m^2)$.

3.6 Clustering

The final stage is to use Quasi-Clique Merge algorithm (QCM [14]) to cluster all documents. We suppose h is the level number of the hierarchical system. Then, by the estimation in [14], the number of iterations is bounded by $O(hn^2log(n))$. Note that, for an input set of n documents, the number of hierarchical levels is $\log(n)$ in average. Thus, the complexity of QCM is $O([nlog(n)]^2)$.

3.7 Total Complexity

By summing up all steps, the total complexity is

$$O(t + m^2\phi^2 + nm^2 + n^2m^2 + [nlog(n)]^2),$$

where $|\Sigma|$ is the number of the distinct alphabets appearing in the key words, t is the average length of the documents, ϕ is the average appearance of a keyword in a document, m is the total number of keywords, n is the total number of documents.

Since we compare lots of documents, so ϕ (the average appearance of a keyword in a document) is much smaller than n (the total number of documents), and t is usually less than n^2m^2. Thus, the complexity is further simplified as $O(n^2m^2 + [nlog(n)]^2)$.

4 Experimental Results

In order to evaluate the effectiveness of our algorithm, we will compare the results of our method with the usual keyword frequency method. We calculate the similarity between every pair of documents the following two different ways:

KF method: *only use keyword frequency information.*
KFP method: *use keyword frequency and structure information, which is based on the weighted directed multigraph model described in this paper.*

In the following analysis we will show that the KFP method is superior to the KF method.

4.1 Nigerian Fraud Emails

We acquired 542 different Nigerian Fraud Emails from an internet archive [26]. We wish to cluster these emails in order to determine any commonality in the authorship of the texts.

In the following experiment, we choose {*bank, account, money, fund, business, transaction*} as the keyword set. Consider two emails: *2001-10-11.html*, *2002-08-27.html* (Figure 5).

2001-10-11.html
Sir,

May I briefly introduce myself. I am Edward Mambo of the Democratic Republic of Congo (formerly Zaire) and the former aide de camp of the Late President Laurel Kabila of blessed memory.
I am presently living in Nigeria on asylum. In case you are wondering how I got your contact, I got your contact from a source at the foreign trade office of the Nigeria Chamber of commerce, trade and industry. The events of the past one year in my country has been very unfortunate. The Late president Kabila successfully toppled the very corrupt government of the Late Dictator Mobutu Sese Seko and ruled justly and peacefully for a few months. Unfortunately, the Tutsi from Goma, the Northern part of the country rebelled against his government and since then my country has been at war.
I am very dedicated and committed to winning the war against the rebels until recently when to my shock I found out that Senior Army Officers and Government officials were stealing public funds and looting government treasury and sending them to foreign countries. They have exploited the war situation to bring back the lawlessness and corruption of the Mobutu days.
After the late President Kabila was assainated in very questionable circumstances, I knew I was in danger of losing from both ends. Due to this development and as a way of survival, I have joined the train. I have in my possession funds amounting to US$15.5million. In the usual manner of conveying such funds, the money is in defaced form in a trunk box for security reasons. It is now in the safe custody of a security company. The company is however not aware of the real contentsof the trunk box.
My problem is that the financial Law of Nigeria does not give assylum seekers any financial rights. And because of the economic instability and religious crisis, I have decided that it will be unwise to attempt investing the funds here. The climate here is very unfavourable and as a result, I have not disclosed even to my hosts anything about this fund.
Now I am seeking foreign assistance to transfer the funds overseas. If you can assist, I am willing to give you 20% of the funds, that is US$3.1Million. You will understand that my entire life and future depend on this money and I shall be very grateful if you can assist me. The major thing I demand from you is the absolute assurance that the money will be safe and you will not sit on it when it is transferred to your account.
If you are ready to be of assistance, please contact me immediately by e-mail, so that I can furnish you with the modalities for the transaction and what is expected of you. I shall be most grateful if you maintain utmost confidentiality and keep this message entirely to yourself.

Best Regards,
MR. Edward Mambo.
234-80-330-46881
Email:edwardmambo@eudoramail.com

2002-08-27.html
STRICTLY CONFIDENTIAL

May I briefly introduce myself. I am DANGO SAKUSHI of the Democratic Republic of Congo (formerly Zaire) and the former aide de camp of the Late President Laurel Kabila of blessed memory.
I am presently living in Nigeria on asylum. In case you are wondering how I got your contact, I got your contact from a source at my country Embassy in Nigeria.
The events of the past two to three years in my country has been very unfortunate. The Late president Kabila successfully toppled the very corrupt government of the Late Dictator Mobutu Sese Seko and ruled justly and peacefully for a few months. Unfortunately, the Tutsi from Goma, the Northern part of the country rebelled against his government and since then my country has been at war.
I am very dedicated and committed to winning the war against the rebels until recently when to my shock I found out that Senior Army Officers and Government officials were stealing public funds and looting government treasury and sending them to foreign countries. They have exploited the war situation to bring back the lawlessness and corruption of the Mobutu days.
After the late President Kabila was assainated in very questionable circumstances, I knew I was in danger of losing from both ends. Due to this development and as a way of survival, I have joined the train. I have in my possession funds amounting to US$15.5million. In the usual manner of conveying such funds, the money is in defaced form in a trunk box for security reasons. It is now in the safe custody of a security company. The company is however not aware of the real contents of the trunk box.
My problem is that the financial Law of Nigeria does not give asylum seekers any financial rights. And because of the economic instability and religious crisis, I have decided that it will be unwise to attempt investing the funds here. The climate here is very unfavourable and as a result, I have not disclosed even to my hosts anything about this fund.
Now I am seeking foreign assistance to transfer the funds overseas. If you can assist, I am willing to give you 20% of the funds, that is US$3.1Million. You will understand that my entire life and future depend on this money and I shall be very grateful if you can assist me. The major thing I demand from you is the absolute assurance that the money will be safe and you will not sit on it when it is transfered to your account.
If you are ready to be of assistance, please contact me immediately by e-mail, so that I can furnish you with the modalities for the transaction and what is expected of you. I shall be most grateful if you maintain utmost confidentiality and keep this message entirely to yourself.
Please send your response to sakushi@lycos.com. My mobile line is 234-803-3297612 we are 1hr ahead GMT (London) time.

Best Regards,
DR. DANGO SAKUSHI.

Fig. 5 Emails: 2001-10-11.html and 2002-08-27.html

The similarity between these two emails via the KF method is 1; the similarity between these two emails via the KFP method is 0.999992. Reading both emails shows that they are almost the same. For these two emails, both algorithms provided proper estimation of their similarity.

This does not hold in general, for the following example shows a "false positive" output by KF method. Consider the pair of emails: *2002-02-20a.html*, *2002-07-04b.html* (Figure 6).

Inspection of the documents clearly shows that they are written in very different styles. The similarities estimated by KF and KFP methods are 1 and 0.43177, respectively. It is evident that one is not able to distinguish these two emails by KF method, while the estimation of similarity by KFP method is much more reasonable.

2002-02-20a.html

Calvary Greetings.

I am a preacher with the Seed Harvest Ministry, and I will like to get advice from you. I believe the advice I need is secular to some extent, but do have the patience to understand my intention.
A few year's ago I was in Liberia where I had established a little congregation where I preached regularly, but the civil war escalated and the church was converted to a hospital of sought. On one faithfull day three Nigerian Soldiers came to me and left trunks of money with me and swore to come back for it.
When the war in Liberia intensified I deposited this box in a security vault in The Netherlands (Holland), I was escorted to Holland, by the ECOMOG soldiers who felt the content of the trunks were documents and never cared to check the trunks, I was air lifted to England with other refugees then, I stayed in England and when I was getting frustrated I returned to Africa only to find out that two of the three Soldiers died in Battle while the last was declared MIA (Missing in Action).
Now I have the fund, which the soldiers said was over thirty million dollars (US$30 million), but it lies in a security vault in a Security Company in Holland. I was contacted in a letter dated 12th of September by the security company asking me to indicate if I was in receipt of the letter as a sign that I am living and still subscribed for their service.
I am interested in using a small fraction of this money, much less than one percent for a re- organization of the work of God, but I do not need the rest and do not want to have any direct dealing with it, but I need someone who will be able to use the fund better maybe for charity or something universally profitable, I have thought of doing it myself but, my ministry is the apocalypse and I believe and preach the soon coming of the Lord which make me not indulgent in reliance on money or wealth in any form.
I also do not have the money that has accrued on the service and safe deposit of the company that holds the fund, I will be glad if you have any interest in the direction of maybe managing or transmitting the fund as you prefer in your capacity and understanding.
I will be glad to get a response from you ASAP.

Regards.
Rev. Sunday Nana

2002-07-04b.html

Pardon the abruptness and the liberty of this letter; It is due to its exigency.
I am a government functionary in my country. I Represent a caucus in need of a foreign partner to assist in the transfer and subsequent investment of a confidential sum in confidence.
We are ready to offer you a percentage for your assistance in the transfer and a negotiable fee if you can put together a portfolio for investment.
All we need is your trustworthiness and your capability to protect our interest.
Please note that we shall go through the legal procedures entailed in our laws and international laws in transferring the funds to you in trust. The source of the money is authoritative.
I wish that you would take the essence of this Letter in strict confidence.
You can reach me immediately for us to commence dialogue on the modi operandi entailed first, by email in earnest. Thank you for your courtesy.

Best regard,
Franklin Numa.

Fig. 6 Emails: 2002-02-20a.html and 2002-07-04b.html

4.2 Plagiarism Papers

Plagiarism in academic articles is a well-known issue. The widespread use of computers and the Internet has made it easier to plagiarize the work of others. Most cases of plagiarism are found in academia, where documents are typically scientific papers, essays or reports [27]. Our experiments show that the KFP method can be used to detect the plagiarism very efficiently.

In this case study our methodology involved the acquisition of a well-known plagiarised paper [28] (named *Paper-1A*) on the independence number of a graph and its corresponding original paper (named *Paper-1B*). In order to test whether our algorithm can detect the plagiarism, we randomly download a set of another 35 academic papers from the internet (named *Paper-2, Paper-3, ... , Paper-36*), which are all related to the same subject, that is the independence numbers of graphs. Figure 7 is the first pair of papers: Paper-1A and Paper-1B.

All of the papers are obtained as *pdf* files. Due to the limitation of the technology, when we convert those pdf files into text files, mathematical formulas are not able to be converted in a proper way: the same formula from different pdf files may converted into very different sequences consisting of special symbols separating with various number of spaces. It will definitely introduce errors when calculating the distance between keywords. In order to eliminate the errors introduced when converting the pdf files into text files, we will use the number of alphabets between the keywords (instead of the number of words between keywords) as the distance between keywords.

The keywords set consists of 23 frequently used terminologies in graph theory. Table 4 and Table 5 indicate the significant difference in the applications of both methods: KF and KFP.

From Table 4, estimated by KFP method, the similarity between the Paper-1A (the plagiarism paper) and Paper-1B (the original paper) is 0.78, and the similarities between all other pairs of papers are less than 0.6, most

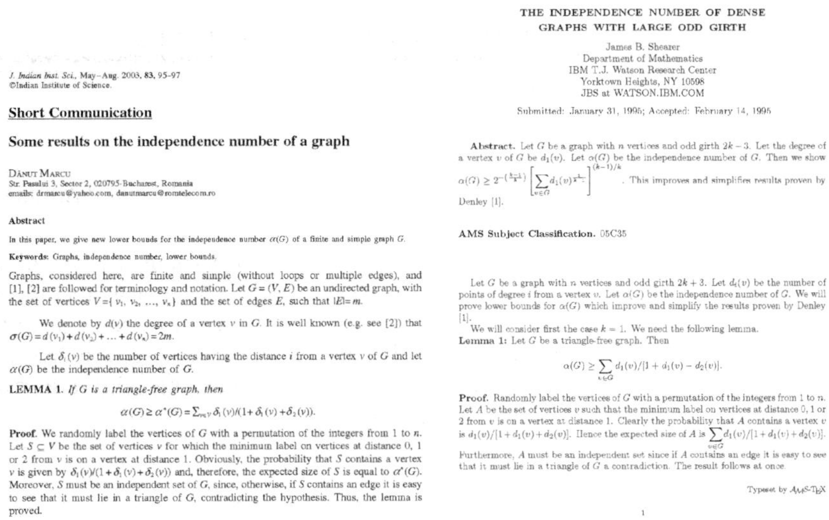

J. Indian Inst. Sci., May–Aug. 2003, 83, 95–97
©Indian Institute of Science.

Short Communication

Some results on the independence number of a graph

DĂNUŢ MARCU
Str. Pasului 3, Sector 2, 020795-Bucharest, Romania
emails: drmarcu@yahoo.com, danutmarcu@romtelecom.ro

Abstract

In this paper, we give new lower bounds for the independence number $\alpha(G)$ of a finite and simple graph G.

Keywords: Graphs, independence number, lower bounds.

Graphs, considered here, are finite and simple (without loops or multiple edges), and [1], [2] are followed for terminology and notation. Let $G = (V, E)$ be an undirected graph, with the set of vertices $V = \{v_1, v_2, \ldots, v_n\}$ and the set of edges E, such that $|E| = m$.

We denote by $d(v)$ the degree of a vertex v in G. It is well known (e.g. see [2]) that $\sigma(G) = d(v_1) + d(v_2) + \ldots + d(v_n) = 2m$.

Let $\delta_i(v)$ be the number of vertices having the distance i from a vertex v of G and let $\alpha(G)$ be the independence number of G.

LEMMA 1. *If G is a triangle-free graph, then*

$$\alpha(G) \geq \alpha^*(G) = \textstyle\sum_{v \in V} \delta_1(v)/(1 + \delta_1(v) + \delta_2(v)).$$

Proof. We randomly label the vertices of G with a permutation of the integers from 1 to n. Let $S \subseteq V$ be the set of vertices v for which the minimum label on vertices at distance 0, 1 or 2 from v is on a vertex at distance 1. Obviously, the probability that S contains a vertex v is given by $\delta_1(v)/(1 + \delta_1(v) + \delta_2(v))$ and, therefore, the expected size of S is equal to $\alpha^*(G)$. Moreover, S must be an independent set of G, since, otherwise, if S contains an edge it is easy to see that it must lie in a triangle of G, contradicting the hypothesis. Thus, the lemma is proved.

THE INDEPENDENCE NUMBER OF DENSE GRAPHS WITH LARGE ODD GIRTH

James B. Shearer
Department of Mathematics
IBM T.J. Watson Research Center
Yorktown Heights, NY 10598
JBS at WATSON.IBM.COM

Submitted: January 31, 1995; Accepted: February 14, 1995

Abstract. Let G be a graph with n vertices and odd girth $2k - 3$. Let the degree of a vertex v of G be $d_1(v)$. Let $\alpha(G)$ be the independence number of G. Then we show $\alpha(G) \geq 2^{-\left(\frac{k-1}{k}\right)} \left[\sum_{v \in G} d_1(v)^{\frac{1}{k-1}}\right]^{(k-1)/k}$. This improves and simplifies results proven by Denley [1].

AMS Subject Classification. 05C35

Let G be a graph with n vertices and odd girth $2k + 3$. Let $d_i(v)$ be the number of points of degree i from a vertex v. Let $\alpha(G)$ be the independence number of G. We will prove lower bounds for $\alpha(G)$ which improve and simplify the results proven by Denley [1].

We will consider first the case $k = 1$. We need the following lemma.

Lemma 1: Let G be a triangle-free graph. Then

$$\alpha(G) \geq \sum_{v \in G} d_1(v)/[1 + d_1(v) - d_2(v)].$$

Proof. Randomly label the vertices of G with a permutation of the integers from 1 to n. Let A be the set of vertices v such that the minimum label on vertices at distance 0, 1 or 2 from v is on a vertex at distance 1. Clearly the probability that A contains a vertex v is $d_1(v)/[1 + d_1(v) + d_2(v)]$. Hence the expected size of A is $\sum_{v \in G} d_1(v)/[1 + d_1(v) + d_2(v)]$. Furthermore, A must be an independent set since if A contains an edge it is easy to see that it must lie in a triangle of G a contradiction. The result follows at once.

Typeset by $\mathcal{A}\mathcal{M}\mathcal{S}$-TEX

1

Fig. 7 Paper-1A and Paper-1B

Table 4 Similarity Comparison 1

	KFP Method	KF Method
Similarity between Paper-1A and Paper-1B	0.778566	0.97074
Similarity between other pairs of papers	1. All less than 0.6. 2. Most of them are far less than 0.2.	6 pairs of papers have similarities greater than 0.97.

Table 5 Similarity Comparison 2

		Similarity By KFP Method	Similarity By KF Method
Paper-1A	Paper-1B	***0.778566***	0.97074
Paper-25	Paper-34	0.345626	***0.994996***
Paper-21	Paper-34	0.203773	0.985672
Paper-13	Paper-25	0.098588	0.980111
Paper-7	Paper-16	0.077647	0.973067
Paper-16	Paper-23	0.055026	0.971901

of them are far less than 0.2. This strongly indicates that the KFP method works perfectly for the detection of a plagiarism paper.

However, if we use KF method, the similarity between the plagiarism paper and the original paper is 0.97074(see Table 5). And we also find other 6 pairs of papers have similarities greater than the similarity between the plagiarism paper and the original paper. For example, the similarity between Paper-25 and Paper-34 is above 0.99, (note that the similarity between these two papers by KFP method is 0.35). From Table 5, we can see that KFP method performs better than KF method.

5 Conclusions and Future Work

In this paper, we introduced a weighted directed multigraph to model a text document. This method considers not only the keyword frequency information, but also the structure information in the form of the relations between keywords in documents. Through experiments performed on a set of emails and a set of research papers on graph theory, it is evident that the weighted directed multigraph model achieves significantly better than the commonly used frequency only model.

We performed experiments on two sets of documents. For the set of graph theory publications, publicly accessible knowledge about identified plagiarised papers provides us a meaningful "yardstick" for the measurement of the accuracy and effectiveness of our novel method. We may summarize our result with the following conclusion: *the KFP method is able to single out the plagiarised pair with the highest similarity which is much larger than any other pair of papers, while the KF method produces may results without any meaningful gap of similarity to distinguish positive and negative results.*

We also tried a weighted undirected multigraph model (i.e, neglect the direction from one keyword to the other keyword in the graph). Although it will lose some structure information of the document, the result is also very similar to what we described above. The advantage of undirected version is the significant reduction of the usage of memory space comparing with the weighted directed multigraph model.

These initial results indicated that the algorithm is much more effective at discriminating and clustering text documents and further improvement of accuracy and performance is expected. Specificially, it is anticipated that one can construct an ontological representation of the semantic information [9, 17, 2] to further enhance the KFP measure and that this information can then be used to set up the directed weighted multigraph. This will in turn allow us to use QCM method to classify all documents with even better precision.

Representing a document as a weighted directed multigraph model is the novel idea introduced in this paper. This approach enables us to further distinguish documents from the SAME category into smaller groups base on writing style, or subcategory. We also believe this weighted directed multi-graph model has a great potential to be applied to other data mining research in information related fields.

Acknowledgements. This work was supported in part by a WV EPSCoR Research Challenge Grant.

References

1. Apte, C., Damerau, F., Weiss, S.: Text mining with decision rules and decision trees. In: Workshop on Learning from text and the Web, Conference on Automated Learning and Discovery (1998)
2. Bestgen, Y.: Improving Text Segmentation Using Latent Semantic Analysis: A Reanalysis of Choi, Wiemer-Hastings, and Moore. Computational Linguistics 32(3), 455 (2006)
3. Hansen, P., Jaumard, B.: Cluster analysis and mathematical programming. Mathematical Programming, 191–215 (1997)
4. Hardle, W., Simar, L.: Applied Multivariate Statistical Analysis. Springer, Berlin (2003)
5. Hassan, S., Mihalcea, R., Banea, C.: Random-Walk Term Weighting for Improved Text Classification. In: Proceedings of the IEEE International Conference on Semantic Computing (ICSC 2007), Irvine, CA (September 2007)
6. Jackson, P., Moulinier, I.: Natural Language Processing for Online Applications: Text Retrieval, Extraction, and Categorization. John Benjamins Publishing Co., Amsterdam (2002)
7. Joachims, T.: Text categorization with support vector machines: Learning with many relevant features. In: Nédellec, C., Rouveirol, C. (eds.) ECML 1998. LNCS, vol. 1398, pp. 137–142. Springer, Heidelberg (1998)
8. Lan, M., Tan, C., Low, H., Sungy, S.: A comprehensive comparative study on term weighting schemes for text categorization with support vector machines. In: Proceedings of the 14th international conference on World Wide Web, pp. 1032–1033 (2005)
9. Landauer, T.K., Foltz, P., Laham, D.: An introduction to latent semantic analysis. Discourse Processes 25, 259–284 (1998)
10. Lewis, D.D.: Naive (Bayes) at forty: The independence assumption in information retrieval. In: Nédellec, C., Rouveirol, C. (eds.) ECML 1998. LNCS, vol. 1398, pp. 4–15. Springer, Heidelberg (1998)
11. Milligan, G.W.: Cluster analysis. In: Kotz, S. (ed.) Encyclopedia of Statistical Sciences, pp. 120–125. Wiley, New York (1998)
12. Mitchell, T.: Machine Learning. McGraw-Hill, New York (1997)
13. Ng, H., Goh, W., Low, K.: Feature selection, perceptron learning, and a usability case study for text categorization. In: Proc. 20th Int. ACM SIGIR Conf. on Research and Development in Information Retrieval (SIGIR 1997), pp. 67–73 (1997)
14. Ou, Y., Zhang, C.-Q.: A new multimembership clustering method. Journal of Industrial and Management Optimization 3(4), 619–624 (2007)
15. Ponte, J.M., Croft, W.B.: A language modeling approach to information retrieval. Research and Development in Information Retrieval, pp. 275–281 (1998)

16. Robertson, R., Sparck-Jones, K.: Simple, proven approaches to text retrieval. Technical Report (1997)
17. Rosario, B.: Latent Semantic Indexing: An overview. INFOSYS 240 (Spring 2000)
18. Ruiz, M.E., Srinivasan, P.: Hierarchical text categorization using neural networks. Information Retrieval 5(1), 87–118 (2002)
19. Schutze, H., Hull, D.A., Pedersen, J.O.: A comparison of classifiers and document representations for the routing problem. In: Proceedings of the 18th annual international ACM SIGIR conference on Research and development in information retrieval, Seattle, Washington (1995)
20. Sebastiani, F.: Machine learning in automated text categorization. ACM Computing Surveys 34(1), 1–47 (2002)
21. Tong, S., Koller, D.: Support Vector Machine Active Learning with Applications to Text Classification. Journal of Machine Learning Research 2, 45–66 (2001)
22. de Vel, O., Anderson, A., Corney, M., Mohay, G.: Mining E-Mail Content for Author Identification Forensics. SIGMOD Record 30(4), 55–64 (2001)
23. Xu, Y., Olman, V., Xu, D.: Clustering gene expression data using graph-theoretic approach: an application of minimum spanning trees. Bioinformatics 18, 536–545 (2002)
24. Yang, Y., Liu, X.: A re-examination of text categorisation methods. In: Proc. 22nd Int. ACM SIGIR Conf. on Research and Development in Information Retrieval (SIGIR 1999), pp. 67–73 (1999)
25. Yang, Y., Pedersen, J.O.: A comparative study on feature selection in text categorization. In: Proceedings of the 14th International Conference on Machine Learning, Nashville, US (1997)
26. Nigerian Fraud Email Gallery, http://potifos.com/fraud/
27. http://en.wikipedia.org/wiki/Plagiarism
28. http://en.wikipedia.org/wiki/D%C4%83nu%C5%A3_Marcu

Retrieving Wiki Content Using an Ontology

Carlos Miguel Tobar, Alessandro Santos Germer, Juan Manuel Adán-Coello, and Ricardo Luís de Freitas

Abstract. This chapter addresses a question regarding relevant information in a social media such as a wiki that can contain huge amount of text, written in slang or in natural language, without necessarily observing a fixed terminology set. This text could not always be adherent to the discussed subject. The main motivation leads to the need of developing methods that would allow the extraction of relevant information in such scenario. A result system was designed upon ideas from the semantic Web combined with adaptive mechanisms and a modification of the classic vector model for information retrieval. The semantic information is not embedded in the media but within a structurally independent ontology. It was implemented using Java and a MySQL database. The objective was the achievement of, at least, 80% for recall and precision on the system results. The system was considered successful by achieving rates of 100% of recall and approximately 93% of precision.

1 Introduction

Social media can be understood as an on-line, in-the-Web, communication service for humans. According to Mayfield [1], a social media can share: participation, openness, conversation, community, and connectedness. A wiki is a social media that allows people to edit content in a collaborative fashion.

Wikis can be private or open, and are used as informational resources for discussion, as a portal, maintained by a community. One such of these is the wiki for the Fedora Marketing Project [2].

Usually, wikis contain ever increasing huge amounts of text data, which are inserted using natural language, without concerns with linguistic formalities, such as spelling. They even can contain slang as well as bad formed words and expressions.

A wiki usually focuses an area of interest or a specific discussion subject. In the case of general focus, wikis can be used for mining potential customers or information on customer satisfaction among several other issues. Even with an established focus, individuals usually insert not related text in a wiki.

Carlos Miguel Tobar · Alessandro Santos Germer · Juan Manuel Adán-Coello ·
Ricardo Luís de Freitas
Pontifical Catholic University of Campinas

I.-H. Ting et al. (Eds.): Mining and Analyzing Social Networks, SCI 288, pp. 21–33.
springerlink.com

With this scenario in mind, it is extremely desirable the possibility to retrieve just relevant information.

In the next sections are presented several points of an on-going effort to develop methods for extraction of relevant information from wikis in the form of retrieval tools. Next, it is presented the semantic approach that has been used, which is based on an ontology and a classic retrieval model.

The ontology structure and content are discussed in the first main section of the article. The second main section is concerned with the adopted retrieval algorithm. Those two sections are followed by: the presentation of the effort to assess a developed software tool, the discussion of results, the presentation of related work, and the presentation of conclusions.

2 The Semantic Approach

The Web has been a platform for the offer of information objects and information services, including e-commerce systems. The amount of information resources, because of their constant increase in amount and in size, represents a big challenge for the retrieval of relevant information. Information is there to be processed by humans, not by computers.

The Semantic Web (SW) was conceived by Berners-Lee and colleagues [3] to overcome the information overload, in such a way that it should be possible that resources of every type could be localized, retrieved and processed without human intervention, using semantic descriptions amenable to be processed by software agents [4].

Most of the Web sites present a clear separation between author and reader [5]. Authoring tools require special programming skills and, usually, are not collaborative tools. In addition, it is extremely difficult to change or aggregate information to already authored pages. One general solution is the wiki, a service that allows readers change and create new pages in an orderly way, which permits controlled collaboration.

Wikis are social media that can be used to exploit user satisfaction, exchange of interests and experiences, and so on. Through them it is possible to do data mining concerned to tendency information, product acceptability, among others. Without some kind of semantics associated to the wiki content this data mining and even simpler information retrieval is not an easy task.

Following it is briefly presented the concepts that were used to develop a tool that retrieves information from wikis using ideas from the semantic Web. In addition, it is presented one chosen form to determine semantic similarity of documents, which is needed to retrieve general information.

2.1 The Semantic Web

Users basically have two ways to find the information for which they are interested in: they can browse or they can use a search engine [6]. Browse can be time consuming and is prone to distractions. Search engines, based on keywords, usually retrieve large amounts of documents. Among those are irrelevant documents, which have to be discarded by the user without support or with very limited support of automated tools. Moreover, the dynamic nature of the Web requires that the user periodically repeats this search-retrieve-filter process to localize new resources of interest and to update previous ones [7].

As the Web is an ever increasing huge information space, a precise search engine, if existent, is not enough for several user requirements, such as updated information from economic news or shoppers´ feedback. Beyond accurate and standardized metadata describing relevant pieces of information, information agents that continually browse the Web are necessary to search for updated resources of interest.

Most of the existent documents in the Web present a multimedia, unstructured or semi-structured nature. Some are rendered dynamically, meaning that their content is stored inside databases, in which case the content that is located automatically by information agents in the Web are just links and scripts.

The SW can be represented by a stack of systems with seven layers. Basically the base layers are already well specified and consist of systems to process representation schemes for character and resource identification. The following two layers were developed afterwards the appearance of the Web and are concerned with metadata languages for resource description, and with semantic statements about described resources. All these four layers have standard and consolidated specifications, which can evolve and are discussed within the W3C forum [8]. The upper four layers are subject of research, application demonstration construction, and standard submission. They are concerned with: the representation of information on object categories and how objects are interrelated, named ontology; the examination of different ontologies to find new relations among terms and data in them, according to rules defined to be inferred; and the extent to which the information found is both accurate and trustworthy, once machines should be able to discover relevant and quality content more efficiently.

2.2 Information Retrieval in Wikis

In semantic wikis users edit pages with links forming a network that can be queried. They, in different ways, try to offer a kind of semantic search promised by the mentors of the SW [5]. It is interesting to highlight that a semantic wiki is a semantic artifact that allows the creation of content semantically enriched, preferably without a design process beforehand.

All of the known research efforts to create semantic wikis, aggregated in their kernel, explore semantic representations of information objects through the use of ontologies.

There is a semantic engine that explores: in-line annotations (metadata), inserted during the authoring of the wiki content; or metadata that replaces the wiki content. These metadata allows the wiki platform to establish and organize inter related concepts and attributes, which would represent the wiki domain of interest.

Shawn [9], SemperWiki [10], Kaukolu [11], Makna [12], Semantic Media Wiki [13]and WikSar [14] are examples semantic wikis with in-line annotation insertion.

OntoWiki [15], OpenRecord [16] and SweetWiki [17] are examples of wikis where content is replaced by metadata.

2.3 The Adopted Approach

The adopted approach described in this chapter for semantic retrieval is based on two main resources: an ontology and an algorithm derived from the classic vector model [18].

An ontology is a means to formally model a domain of interest [19]. It consists of a hierarchy of concepts, named classes, and associations between concepts. Both classes and associations can be instantiated.

There are not so many useful ontologies as the research community would like [20], because of unresolved technical limitations or the inexistence of sound rationales for why individuals refrain from building them; maybe due to the fact that an ontology should result from the common understanding of a community on the domain of interest.

Traditionally, an ontology follows a social concept of being the result of an agreement on the understanding of a domain by some community.

Although, it is originally intended to make an explicit commitment to shared meaning among an interested community, individuals can use ontologies to describe their own data [4]. For the considered approach, the ontology is not a mechanism for communication and understanding of different agents but an adaptive information mechanism for a specific enterprise. Even though, a common ontology could be used as fine as a personal one.

The intended ontology represents the specific perspective of an organization, on the domain where the wiki subject belongs regarding its interests. The kind of adaptive behavior that is intended is also named personalization, when a determine user is considered during her interactivity with the system.

Most systems with personalized behavior are based on some type of user profile, such as an ontology in the OBIWAN project [6] and in the MESH project [21]. The intended ontology is also used for adaptive behavior, without a specific user, and should be build by the organization carefully, although explicit profile creation is not recommended to avoid a burden on an individual user.

Ontologies enable the formalization of preferences in a common underlying, interoperable representation, where interests can be matched to content meaning [21].

Differently from semantic wikis, the adopted approach is intended to extract information from plain wikis in a semantic way based on an ontology. The tool described in this chapter performs retrievals from already edited wikis. There are no changes in the way wikis are edited. There is no need to edit annotations and include them in text passages.

For ontology representation, it was chosen the Ontology Web Language (OWL) [22], based on the Resource Description Framework (RDF), and it was used Protégé [23] as an ontology editor.

In the implemented tool, considering a wiki content, it is possible to process slang, along with not formal and well written text. Comparable efforts do not support such functionality, as far as it is known.

A *topic* in a wiki corresponds to a document. It is composed by an article and a discussion. Regarding relevance, a calculation is done for the article and other for the correspondent discussion. The higher relevance grade of both is considered to be the topic relevance.

In order to decide if topic part is relevant, the information retrieval vector model is used. Considering a set of documents and a query to retrieve the most relevant documents, each document is represented as a vector. Each vector element corresponds to a separate term in the document set upon which the query is performed. If a term occurs in a specific document, its value in the corresponding vector is non-zero. There are different ways of computing term values, also known as term weights. Considering n as the amount of terms in a vector of a specific document set, each vector can be seen as a point in a n-dimensional space. Similarly a vector is defined for the query, as it was a document. The similarity of one document and the query can be measured by the distance of their correspondent n-space points.

The information retrieval vector model is in essential a classification model that allows handling large volumes of data [6].

The mathematic formulas of the classic vector model were modified in order to consider the semantic nature of the ontology elements, such as the case of a concept that is related to other concepts. The relevance of several associated semantic terms should be weighted higher than the relevance of a isolated one.

3 Ontology Definition

A tool was designed to retrieve the recent and relevant content of a wiki, whose location should be informed together with the ontology to be considered.

Using OWL, *Annotation* and *Object Properties*, synonyms, related verbs, and relevance weighting adjustments can be incorporated into each of the ontology classes in order to allow the calculation of relevance grades.

3.1 Classes and Instances

The Protégé editor allows the creation and maintenance of classes, subclasses and instances in an ontology. Class names should be keywords that reference main concepts in the domain of interest.

Relevance grades for a given class are calculated according to the depth of the class in the ontology hierarchy. For instance, in a three level class hierarchy, a first level class receives a 0.33 relevance weight. A second level class receives a 0.66 and the leaf classes in the hierarchy tree receive a 1.

The developed tool, when querying classes, considers composed words through underscore identification or through the CamelBackCase syntax.

3.2 Annotation Properties

Three annotation properties were chosen to extend the semantic meaning of classes or instances in an OWL ontology. Their meanings are:

- A **synonym** allows the insertion of a word or a list of words that present the same meaning as the class name. The same relevance weight in the ontology for the class name is adopted for its synonyms during relevance calculations.
- A **related verb** offers a similar meaning of a synonym except that the developed tool would try to obtain radicals for the listed verbs [24], which allow a broader search for terms that represent one of the different verb conjugations.
- A **relevance adjustment** is a specified value ($0 \leq v \leq 1$) that substitutes the default value that considers the hierarchy depth of a given class.

3.3 Object Properties

An object property allows interrelating classes or instances. In the relevance algorithm for classes, the existence of such property is interpreted as the existence of a cohesion grade between class families, independently of which class in the family is involved in the property. An object property specifies that it is really relevant if terms of both involved families are present in a given text.

It is possible to have an inverse object property, i.e., the existence of an object property from an origin to a destiny together with another from the previous destiny to the previous origin. For the retrieval algorithm this does not result in a double relevance grade, just one.

4 The Information Retrieval Algorithm

The classic vector model was chosen because it provides a good computational processing level combined to satisfactory results for most of known retrieval solutions. Even though, modifications were done to adequate it to the desired semantic contextualization.

In the following subsections are described the implemented modifications to the original model together with their motivations and the mathematic impacts that they produce.

4.1 Multiple Query Vectors

Each of the documents in a collection has its own term vector for which the distance is calculated regarding the term vector of a query or of an example document, which is the base for retrieval. For the semantic approach, each class family in an ontology gives origin to a term vector, *i.e.*, the ontology can produce several vectors.

In addition regarding an ontology, a term equivalent is a set with the class or instance name, and all synonyms and related verbs that are present in the annotation properties.

While, in the original model, term weights are equal to the number of occurrences of a given term in one of the documents to be searched, in the semantic approach each occurrence of the name, or of each synonym or of each related verb contributes with one to the weight in the term vector of a wiki element.

4.2 Universe of Analyzed Documents

Through the interface of the retrieval tool it is possible to configure where in the wiki site topics should be considered in a retrieval process. It is done through the selection of what wiki URLs are considered more interesting. A drag-and-drop mechanism is used. The user chooses the wiki and browses its content. If one page is considered interesting, the user selects its URL and drag it to an icon that represents the retrieval tool. The icon was drawn to look like a dark hole.

The possibility of considering a whole topic part or just the latest revisions on them allows knowing the recent contributions concerning a given time period. To decide what topics or part of them are recent, the creation and modification dates are stored along with the correspondent topic part in a database.

Each wiki topic is equivalent to two queries in the classic vector model and a distance is calculated concerning each different class family in the ontology. Each ontological family is equivalent to one document in the collection considered in the classical model.

4.3 Semantic Weight

The new semantic scenario requires a new weighting scheme to quantify the relevance of term equivalents (ontology concepts elements) in each class family against each considered topic part. The new weight is named *semantic weight* and is calculated according to the location of each term equivalent in the hierarchy of each class family or through the Relevancy Adjust property. Considering k as the kth concept in the equivalent to a term vector, its depth in a class family cf_k where it appears, and the greatest depth among the ontology class families $maxdepth_{cfk}$ the semantic weight sw_k formula for k is presented in (1).

$$sw_k = (depth_{k,\ cfk}) / (maxdepth_{cfk}) \tag{1}$$

4.4 Inverse Document Frequency

The inverse document frequency *idf* is an indicator in the vector model that benefits documents with terms whose frequency is relatively low concerning the total document set. It is also responsible to avoid that highly frequent terms influence relevance calculations. In order to avoid the appearance of severe numeric distortions, the logarithm function is used. The original formula to compute de *idf* is shown in (2).

$$idf_k = log(N / n_k) \tag{2}$$

Where N is the amount of elements in the document set and n_k corresponds to the number of documents where the kth term occurs, ignoring the amount of its occurrences in each document.

Considering the formula in (2), it can be perceived a drawback. Relevant terms that appear in all considered documents turn to make no positive influence in the calculation results because the obtained *idf* is zero. Because the *idf* index is used in other formulas as a multiplying factor, correspondent results will be zero.

Although common in all texts, a concept with highly semantic weight should not have a null *idf* value. To avoid this distortion, constants were included in the *idf* formula, to maintain a behavior close to the original without allowing a null result, as can be seen in (3).

$$idf_k = log\ ((N + 2) / (n_k + 1)) \tag{3}$$

The new constants in (3) do not cause a significant difference for those cases that present non-null results in (2). This is obtained due to the expected magnitude of N and n_k, corresponding to a large number of documents in the collection and any frequency of the considered concept.

4.5 Normalized Frequency

For each concept k of each document equivalent (class family - d_j), it is counted the number of occurrences of this concept, $freq_{k,j}$. All the frequencies are normalized to a value between 0 and 1, through the division by $maxword_j$, which is the greatest frequency of all concepts (terms) in the document equivalent. This calculation does not suffer any modification, remaining the same as in the original model, (4).

$$f_{k,\,j} = freq_{k,\,j} / maxword_j \tag{4}$$

The same rationale is used to calculate normalized frequencies of concepts in each document equivalent d_j, for each concept considering all document equivalents D (the entire ontology). This normalization is equal in the original model, (5).

$$f_{k,\,D} = freq_{k,\,D} / maxword_D \tag{5}$$

4.6 Concept Weight in Documents

The classic vector model considers the normalized frequency $f_{k,j}$ and the inverse document frequency idf_k in order to obtain the term weight regarding a specific document, as can be observed in (6).

$$w_{k,\,j} = f_{k,\,j} * idf_k \tag{6}$$

For the semantic approach, it was included the semantic weight as a multiplying factor, (7), in order to influence the result value.

$$w_{k,\,j} = f_{k,\,j} * idf_k * sw_k \tag{7}$$

4.7 Concept Weight for Queries

In the original model, to obtain the term weight for a query, the normalized frequency would need to be calculated over the entire document set, not just over a single document. There is the possibility to insert an extra factor in this calculation. This factor can be seen in (8) with the value of 0.5.

$$w_{k,\,q} = (0.5 + 0.5 * f_{k,\,D}) * idf_k \tag{8}$$

The initial reason for the insertion of that factor is that the result would usually be a low number. Thus, the second instance factor of 0.5 produces a half normalized frequency $f_{k,j}$ and a free half value is added to it. This will increase the value of a term that appears more than once in a query.

There was no real reason to consider a constant value in the term weight formula considering the semantic scenario, because the semantic concepts are obtained from the ontology structure. Keyword repetition inside the ontology does not mean that a concept is more relevant in the domain of interest. What really matters is how concepts appear in the ontology hierarchy and how they relate to others.

The semantic weight should be considered with a greater weight than others factors. Thus, the 0.5 factor was substituted by a new factor based on it, as can be observed in (9).

$$w_{k,\,q} = (sw_k + (1 - sw_k) * f_{i,\,D}) * idf_k * sw_k \tag{9}$$

This modification allows the semantic weight to be higher than the normalized frequency $f_{i,Dj}$ and than the inverse document frequency idf_k, considering the document set, once the relevance value is high.

4.8 Similarity between a Document and a Query

To determine the similarity between two documents or between a query and a document, the classic vector model uses the fact that, when the angle between two vectors is very small, the cosine of these two vectors approaches one. The formula used to calculate the cosine can be seen in (10).

$$sim(d_j, q_{cf}) = (d_j \bullet q_{cf}) \,/\, (d_j \times q_{cf}) \tag{10}$$

For the semantic approach the same formula is used. Because a concept vector is created for each class family and also for each topic part, this similarity calculation should be applied several times (number of class families times the amount of topic parts).

4.9 Considering Object Properties

Object properties establish dependency relations between classes or instances. Each of those will produce for each document equivalent d_j, a new weight factor, which will receive as result the multiplication between the two similarities involved, i.e., similarity between d_j and the first class family q_{cf1} and similarity between d_j and the second class family q_{cf2}, (11).

$$w_{objP(cf\,1,\,cf\,2),\,j} = sim(d_j, q_{cf\,1}) * sim\,(d_j, q_{cf\,2}) \tag{11}$$

4.10 Final Ranking

The final raking defines what topics are more relevant regarding the class families in the ontology. This is done through a normalization of factors to obtain the final relevance grade for each topic part. Arithmetic averages are obtained for similarity grades and for the factor from object properties. A final relevance value is obtained for each topic part through the arithmetic mean between the similarity average and the factor average. It is expected that the greatest final relevance values would come from the topic parts whose contents are semantically closer to the domain of interest that is represented in the ontology.

5 Assessment

A semantic-oriented retrieval tool should be satisfactory to any user. For the developed tool, it was defined that it should present at least 80% of both precision and recall, which are well known metrics usually used to assess retrieval processes and tools [18].

Precision is computed as the ratio between the amount of retrieved documents that are relevant and the total amount of retrieved documents. It indicates the capacity to keep out irrelevant documents from the final result. Recall is the ratio between the amount of retrieved documents that are relevant and the amount of documents that should be retrieved. It represents the capacity to retrieve relevant documents.

To assess the developed tool and its retrieval algorithm, a wiki was populated with topics whose content and the expected retrieved result were previously known, considering the defined ontology.

The wiki subject was concerned with a general issue that has to be discussed by a shopper community. Part of the inserted topics was not related to this subject. Thus, the wiki contained relevant and irrelevant information.

The wiki content was composed by 35 topics, with 14 of these, 40% of the topic parts, highly relevant. The other 60% contained other non-related subjects.

Before beginning a retrieval process, the developed tool was configured. More information on this follows in the discussion section.

6 Discussion

The formulas presented previously can be modified. A configuration facility is present in the developed tool that offers the opportunity to reach better and more stable results, which should be different accordingly to the domain of interest. It also will allow the behavior study of the proposed algorithm.

The authors believe that the main influence to obtain relevant results is due to the magnitude of the proposed relevance indices, as can be perceived next.

6.1 Magnitude of Calculated Relevance Indices

One of the main difficulties on using the retrieval algorithm is to understand what relevance indices obtained in a retrieval process mean.

What threshold value, for an index to define a topic part as relevant or not in the domain of interest, is an open question, which should be analyzed for each domain case. In the developed tool, this threshold can be configured to each domain. Thus, the organization should perform a few initial retrieval experiments and configure empirically this threshold, considering the final results of each retrieval effort, which are sorted and presented with their relevance indices.

Class families and properties in an ontology affect the final calculations. In addition, the number of concepts of a domain could be higher than of others, even for comparable topic sets, concerning relevance distribution. Nevertheless, even if the final results are values between 0 and 1, for ontologies representing the same domain or different ones, they will be specific.

6.2 Discrepant Weights

During the tool assessment, it was possible to observe a false-positive case produced by the retrieval algorithm that is very interesting. A topic part that has no relevance at all, considering the ontology, was one of the firsts in the relevance ranking that was produced.

Analyzing the case, the conclusion was that the problem was due to the presence of one isolated keyword in the topic that was spelled the same way as a concept that was present in the ontology. Coincidently this concept does not appear anywhere else. This case caused the associated *idf* to be very high, which influenced the construction of the weight vector for the document equivalent as well as of the query equivalents vectors. As each concept in the vector represents a coordinate in a multiple space, considering both vectors, the correspondent dimension was discrepant to the other concept dimensions, in such a way that the cosine of the angle formed between the correspondent vectors had a very high value.

This discrepant case points out the necessity to include some kind of treatment in the retrieval algorithm that could avoid highly discrepant weights.

6.3 Distinction between Class Families

In the proposed ontology structure there is no way to specify different weights for different class families. If implemented, such functionality will become very interesting because it is acceptable that each class family represents an information sub domain and thus it can be more or less relevant than the other families.

The inclusion of class weights could aggregate a refinement to the information retrieval mechanism that is used by the implemented tool.

7 Conclusions

In this chapter it is presented an approach to perform semantic information retrieval upon wikis. The idea was to provide a tool to follow up news or participation on consumer discussions. The wiki should contain articles and discussions that are inserted continuously during a time frame, but its ideas can be ported to other social media, such as blogs and discussion lists.

The proposed tool can be used in several other scenarios where information retrieval is necessary or can be used for improvements. The main differential to other similar tools and mechanisms is manifold: the semantic nature of the designed algorithm is based on an ontology that uses OWL format; that ontology is used as an adaptive mechanism, almost similar to ontologies with personalization purposes; retrieved results are quite similar to semantic wiki results, with the difference that the wiki does not need annotations in order to obtain semantic relevant information; slang and synonyms are considered; and it was produced a modified version of the classic vector model for information retrieval.

The obtained assessment results were approximately of 100% for recall and 93% for precision. Even though, looking more closely to these impressive results, some problems were identified and should be solved in the future in order to obtain more stability and robustness in the proposed algorithm.

Although some adjustments are necessary to the proposed algorithm for some domain scenarios, it still can be used with relative success to retrieve recent and relevant information from plain wikis.

The designed architecture places the proposed tool outside the Web and the wiki engine. At the beginning, the retrieval target was the MediaWiki software [25], but it is possible to extend the tool to other wiki software by codifying new browsing mechanisms.

Collaborative communities are increasingly present in the Web and they will play major roles in the Web 2.0, so they need semantic treatment, mainly in a scenario of huge data volume.

References

[1] Mayfield, A.: What is Social Media? iCrossing (2008), `http://www.icrossing.co.uk/fileadmin/uploads/eBooks/What_is_Social_Media_iCrossing_ebook.pdf`

[2] The Fedora Marketing Project, `http://fedoraproject.org/wiki/Marketing`

[3] Berners-Lee, T., Hendler, J., Lassila, O.: The Semantic Web, pp. 34–43. Scientific American (May 2001)

[4] Shadbolt, N., Hall, W., Berners-Lee, T.: The Semantic Web Revisited. IEEE Intelligent Systems, 92–101 (2006)

[5] Miilard, D.E., Bailey, C.P., Boulain, P., et al.: Semantics on demand: Can a Semantic Wiki replace a knowledge base? New Review of Hypermedia and Multimedia 14(1), 95–120 (2008)

[6] Gauch, S., Chaffee, J., Pretschner, A.: Ontology-based personalized search and browsing. Web Intelligence and Agent Systems: An international Journal 1, 219–234 (2003)

[7] Adán-Coello, J., Tobar, C.M., Rosa, J.L.G., Freitas, R.L.: Towards the Educational Semantic Web. In: Mendes Neto, F.M., Braileiro, F.V. (eds.) Advances in Computer-Supported Learning, pp. 145–172. Information Science Publishing, United Kingdom (2007)

[8] W3C, World Wide Web Consortium (2009), `http://www.w3.org/`

[9] Aumuller, D.: SHAWN: Structure Helps a Wiki Navigate. MsC Thesis. University of Leipzig, Germany (2005)

[10] Oren, E., Delbru, R., Möller, K., Völkel, M., Handschuh, S.: Annotation and Navigation in Semantic Wikis. In: Proc. of the First Workshop on Semantic Wikis: From Wiki to Semantics, ESWC (2006)
[11] Kiesel, M.: Kaukolu – Hub of the semantic corporate intranet. In: Workshop: From Wiki to Semantics, ESWC (2006)
[12] Makna, http://www.aps.ag-nbi.de/makna
[13] Vökel, M., Krötzsch, M., Vrandeci, D., et al.: Semantic Wekipedia. In: Proceedings of the 15th International Conference on WWW. ACM Press, New York (2006)
[14] Aumueller, D., Auer, S.: Towards a Semantic Wiki experience – Desktop integration and interactivity in WikSAR. In: Proceedings of 1st Workshop on the Semantic Desktop (2005)
[15] Hepp, M., Bachlechner, D., Siorpaes, K.: OntoWiki: Community-driven Ontology Engineering and Ontology Usage based on Wikis. In: Proceedings of the 2005 International Symposium on Wikia (2005)
[16] OpenRecord, http://openrecord.org
[17] Buffa, M., Gardon, F.: SweetWiki: Semantic web enabled technologies in Wiki. Mainline Group, I3S Labortory, University of Nice (2006)
[18] Baeza, Y.R., Ribeiro Neto, B.: Modern Information Retrieval. ACM Press, New York (1999)
[19] Gruber, T.R.A.: Translation Approach to Portable Ontologies Specifications. Knowledge Acquisition 5(2), 199–220 (1993)
[20] Hepp, M.: Possible Ontologies: How Reality Constrains the Development of Relevant Ontologies. IEEE Internet Computing 11(1), 90–96 (2007)
[21] Cantador, I., Fernández, M., Vallet, D., et al.: A Multi-Purpose Ontology-Based Approach for Personalised Content Filtering and Retrieval. In: Wallace, M., Angelides, M. (eds.) Advances in Semantic Media Adaptation and Personalization. Studies in Computational Intelligence, pp. 25–52. Springer, Heidelberg (2008)
[22] OWL Web Ontology Language Guide, W3C (2004), http://www.w3.org/TR/owl-guide/
[23] Protégé Ontology Editor, Stanford University School of Medicine (2009), http://protege.stanford.edu
[24] Porter, M.F.: Snowball: A language for stemming algorithms (2001), http://snowball.tartarus.org/texts/introduction.html/
[25] MediaWiki Project (2009) MediaWiki.org., http://www.mediawiki.org/wiki/MediaWiki/

Ego-Centric Network Sampling in Viral Marketing Applications

Huaiyu (Harry) Ma, Steven Gustafson, Abha Moitra, and David Bracewell

Abstract. Marketing is most successful when people spread the message within their social network. The Internet can serve as an approximation of the spread of messages, particularly marketing campaigns, to both measure marketing effectiveness and provide data for influencing future efforts. However, to measure the network of web sites spreading the marketing message potentially requires a massive amount of data collection over a long period of time. Additionally, collecting data from the Internet is very noisy and can create a false sense of precision. Therefore, we propose to use ego-centric network sampling to both reduce the amount of data required to collect as well as handle the inherent uncertainty of the data collected. In this the book chapter, we study whether the proposed ego-centric network sampling accurately captures the network structure. We use the Stanford-Berkeley network to show that the approach can capture the underlying structure with a minimal amount of data.

1 Introduction

Viral marketing, or word-of-mouth marketing is successful when people take it upon themselves to spread the "message" within their social network [14], potentially reaching an audience much bigger than what the original marketing budget could have obtained through more traditional means. On the Internet, a key to viral marketing is to get a web site to recommend the target of the marketing message to others, their social network as well as to their readers, who will hopefully spread the message and react positively to the message. While the measurable Internet is only an approximation of the real word-of-mouth spread of the message, it is reasonable to argue that, in some domains, the Internet can serve as a coarse approximation of the real social spread of a message [4]. Additionally, as more people use tools like

Huaiyu (Harry) Ma · Steven Gustafson · Abha Moitra · David Bracewell
GE Global Research, One Research Circle, Niskayuna, NY 12309

I.-H. Ting et al. (Eds.): Mining and Analyzing Social Networks, SCI 288, pp. 35–51.
springerlink.com

Facebook and Myspace, the Internet may become a very good proxy for the true social network that it approximates [10].

It is very crucial to the success of a viral marketing campaign that the message reaches the right audience at the right time – an audience who is willing to spread the message effectively within their social network. They are the "opinion leaders" [19], and finding them is very important. However, viral marketing is not as simple as just looking for opinion leaders who are the "most connected". For example, a blogger that is not widely connected may in fact have a high influence if one of her readers is highly connected. We can think of her as an "advisor to opinion leaders". There are other factors to consider regarding the success of viral marketing, for example the coverage over the entire potential audience by the initial adopters will likely have significant impact on the end result [21].

In order to measure the success of marketing campaigns, and to assist in their planning, it is desirable to capture the audience's network and study the spread of the campaign's message, for example see [15]. Audience networks could mean "friends" on a social networking site like Facebook, or they could mean the dedicated readers of a blog. Likewise, to understand the spread of a campaign or "viral" message, we could look at the message passing within a Facebook community, or we could study the online posting of web sites or blogs. Between these two levels of analysis (the low level person-to-person communication, and the high level web site publishing a communication), our objectives is to study the networks of web sites. Just like person-to-person networks, web sites have the potential to influence people and spread viral messages.

Toward our stated objective, we propose to use in-link (directed hyper-link) networks to help determine the attributes of a web site and measure the effectiveness of viral marketing [13]. We are aware of the opportunistic use of links on the Internet to increase the perceived importance or authority of a site. To remedy the uncertainty and noise introduced when using in-link networks, as well as to remove the need to maintain a massive amount of data on all the links over time, we propose to study the network from the perspective of individual web sites (ego-centric networks) and use a sampling based approach.

Ego-centered networks provide a view of the network from the perspective of a single node (ego) and their connections (alters) [20]. Members of the network are defined by their specific relations with ego. This ego-centric approach is particularly useful when the population is large, or the boundaries of the network population are hard to define [22]. Ego-centric analysis is well suited to the study of how web sites maintain wide-ranging relations on the Internet, since no matter how much data we can collect, completely capturing the online social interactions in real-world will not be feasible.

Social Network Analysis (SNA) is used to maximize the impact of viral marketing [20]. SNA views social relationships in terms of nodes and ties (edges). It examines the structure of social relationships, provides both a visual and a mathematical analysis of the relationships and reveals the hidden relationships between people and the patterns and implications of these relationships. We estimate the global properties of web sites by studying their ego networks. The focus is on estimating the network structures and determining their relationship with the global properties of the web sites.

In this the book chapter we describe our method for creating ego-centric networks for the application of viral marketing. To understand how well our method captures the true underlying network, we use a well-known data set, the Stanford-Berkeley network, to represent the 'true' network. We carry-out several experiments to measure if, on several key metrics, our technique provides good approximations with minimal data collection required. The results indicate that the ego-centric network approach captures the structure of the true network, increases in accuracy with more data, and does better than a baseline approach. Next we describe the ego-centric approach, the experiments and results.

2 Ego-Centric Networks

Understanding the web and its properties has been a hot research topic since its inception [17]. One of the biggest algorithmic challenges is that the size of the web is too huge to grasp its complete picture, hence some sampling procedure is beneficial.

Given a huge web graph, how can we derive a representative sample that preserves the major properties of the original graph? [6] proposed a specific random walk on the (directed) web graph, however it is not clear how many steps are required in order to approximate the equilibrium distribution. [1] proposed asking various search engines for the in-links of a given page in order to sample all adjacent edges of a given page. However, frequently only a subset of all in-links can be found in this way. [8] developed the HITS algorithm, which queries an index-based search engine (for example Google.com) to find web pages related to some query. The resulting web pages are then expanded to include in-links to and out-links from the web pages. Network metrics are then used to assign authority and give prominence to more structurally important web pages.

From the social sciences, a related and interesting approach is snow-ball sampling, see [12] for a recent, Internet-related discussion. In this technique, initial seeds are recruited to report their immediate social network, who are then subsquently recruited. Related work has focused on developing sampling techniques that estimate the true population size accurately and avoid biases, an example being respondent-driven sampling [16].

Many known network sampling algorithms may be applied in our ego-centric framework. The natural questions to ask here are 1) which sampling method to use, 2) how small of a sample size can be used and 3) how to scale up the measurements of the sample in order to get the measurements of the whole graph, if needed. [5] present a nice study on some of the sampling methods in terms of how good they are in addressing the above questions. The methods fall into two categories, sampling by random node/edge selection and sampling by exploration. The latter, like snowball sampling, explores the nodes in a given node's vicinity, which fits into our ego network framework. According to their study, one of the better performing methods, if not always the best under different circumstances, is a sampling approach inspired by the work on temporal graph evolution, called Forest Fire sampling. It 'burns', or keeps, outgoing links and the corresponding nodes with a certain probability. If a link gets burned, the node at the other endpoint gets a chance to burn its own links, and so on recursively building a network out of the initial node, burned edges and their nodes. This model has two parameters: forward and backward burning probabilities. We propose a very similar approach in the following section.

3 Methodology

Our ego network sampling approach is a variation of the Forest Fire algorithm. In our algorithm, we set the backward burning probability to zero, as we only allow a node one opportunity to obtain edges linking in to it, and use the Yahoo! Site Explorer for in-link estimation. The following are the steps,

- Given a web site, get the total number of in-links from Yahoo!.[1]
- We get 100 in-links for the web site from Yahoo! [2].
- We find the unique domains in the in-link list and calculate rate, defined as $r = \#ofUniques/100$.
- We randomly pick n links from the unique in-link list [3], where n is a geometric random number with mean proportional to $log(\#\text{in-links} * r)$.
- Repeat the above steps for nodes of the burned links.
- Stop when it burns R levels deep. We suggest using $R = 3$.

We then repeat the above steps to get a total of three random ego networks for the same web site. We assume that the network structure of the web site stays the same within a short time period. The differences observed in the three networks are due to the randomness of the sampling. The three networks are studied individually and in combination.

Figure 1 shows samples of two generated ego networks. Each row displays three sampled networks for a web site. The absolute position of the same nodes (web sites) in each of these graphs is not fixed. That is, a web site may be represented in each of the three ego-networks, but occur in different locations in the actual network displayed. As we can see from the figure, the three generated networks of a web site are different from each other, but the general patterns are preserved. The upper networks show great interconnectivity, while the lower ones are more star-like.

[1] Yahoo! treats a domain suffixed with and without "www." as two different domains. Our solution to this problem is to get the total number of in-links for both, and use the one with the higher value to create the network.

[2] The Site Explorer ranks the in-links in order of importance. It returns a maximum of 100 in-links. We use the in-links options of "Except from this domain" and "to Entire site" to get an accurate picture of the external links.

[3] By our observation, the Yahoo! In-link distribution is extremely heavy-tailed. We understand that taking logarithms will change the degree distribution of the network, but it is the most suitable approach to downsize the network.

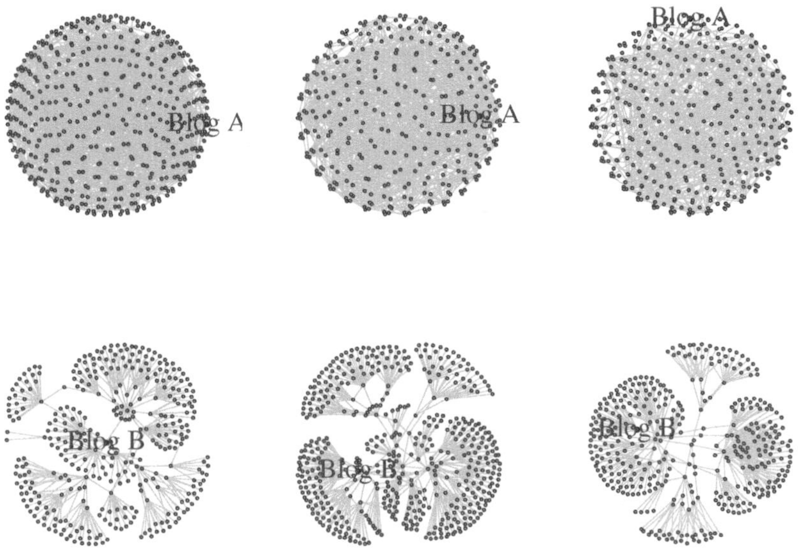

Fig. 1 Sample ego-networks for two web sites, Blog A and Blog B, using the random burn method

4 Network Measures

The goal of this study is to understand if a random network, generated using the best known models of social network growth, can be measured on a node-by-node basis with common network metrics, and have those measurements be similarly captured using our process of creating ego-networks.

There are various centrality measures for finding the important (central) nodes (see [18] for a recent discussion). In this study we utilize the degree distribution and the clustering coefficient to capture the structure of the network. The degree of a node in the network is the number of adjacent edges of the node. The clustering coefficient (CC) measures the probability that the adjacent nodes of a node are connected. It is sometimes also called transitivity. High CC is an indication of a small world, i.e., most nodes can be reached from every other by a small number of steps.

5 Empirical Experiments

We evaluate our approach to find if it can preserve the real network structure and to determine how big of a sample size is needed to achieve good accuracy. The evaluation is done using the Stanford-Berkeley Web network crawled by [7] in December 2002. It contains 685,230 nodes and 8,006,115 links. The connectivity statistics of this network are similar to the results of [2], which shows that the link structure at the level of hosts and domains also follows a inverse power law distribution.

To provide a context for the performance of our approach, we use the Random Edge (RE) sampling approach as the benchmark. The steps of a RE are:

- Start from a web site (ego, or center node), let set $S = \{site\}$;
- Repeat the following, until stopping criteria met
  - Select a node i from S at random;
  - If i is within a certain reach R from $site$ (we use $R = 3$), randomly sample i's edges that have not been previously visited;
  - For each new edge sampled $e_{i,j}$, add j to S.

The stopping criterion we used is the percent of total edges sampled from the known network. We also make sure that the algorithm stops if no edge increases beyond a maximum iteration (we use 100). Our approach for creating ego-networks can be described in a similar fashion as follows:

- Start from a site (ego, or center node) and set $S^0 = \{site\}$;
- Set $R = 3$
- Set $r = 0$
- Repeat the following, until stopping criteria met $(r > R)$
  - Select a node i from S^r at random;
  - Generate a geometric random number g with mean proportional to the number of edges of i and select g edges, where e_{i,j^m} and $m = 1..g$.
  - Add the nodes $j^m, m = 1,...g$ into S^{r+1}.
  - Set $r = r + 1$

We consider the following factors in our experiments:

- Ego Network: We randomly pick eight ego in-link networks ($R = 3$) from the Stanford-Berkeley network[4];
- Sampling Method: We sample the eight ego networks using both the RE method and our method;
- Sample Size: We use random networks of varying sizes. The sample sizes are set to be $\{5, 10, 15, 20, 50, 90\}$ percent of the edges;

We run each experiment 20 times to reduce the simulation error. We study the results from two aspects. Firstly, we want to see whether the sampled ego networks are representative of the true ego networks, and secondly, we want to see whether the sampled ego networks can be used to described the relationship between the

[4] Note that we limit the population to be ego networks that have less than 50,000 edges and a reach (R) bigger than 3. The former constraint is to reduce the computational effort, and the latter is to eliminate simple structured networks, such as star networks.

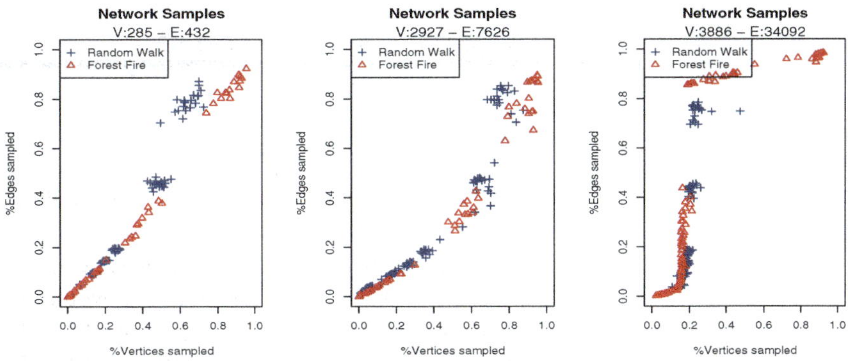

Fig. 2 Network Samples by Different Sampling Methods and Experiment Configurations. "V" and "E" in the subtitles denote the number of vertices and number of edges respectively

different ego networks. Figure 2 illustrates how the resulted networks vary from sample to sample, given different edge sample size configuration, by plotting the percent of edge sampled against percent of vertices sampled for three different networks. We can see the points produced by the RE method produced were clustered around the selected sample size points; while the points yielded by the FF method were more spread out. The FF method sampled more vertices than the RE method when required edge sample size is huge. The RE method more often stopped before reaching the targeted sample size when the required sample size was relatively high.

5.1 Performance Measure

One way of evaluating how good the samples are is to see how good they are in estimating degree distributions. [11] shows that predictions for typical vertex-vertex distance, clustering coefficients and vertex degree based on only degree distributions agree well with empirical data. Figure 3 shows that the eight ego networks selected for our study have different degree distributions: some follow power law and some do not.

The plot on the left in Figure 4 illustrates the estimated degree distributions by the two methods and the true distribution using one of the eight ego networks with a sample size of 5% of the total network. We can see that even though the sampled edges are only about 5% of the total, the degree distribution estimations are not bad for small degrees (< 50). This finding is in line with the study by [9]. They show that only the maximal degrees significantly depend on the sample size and the average degree is roughly constant. The plot on right shows the estimations of the CC distribution conditional on node degree. The dots represent the conditional CC's and the curves are the spline fits. Both methods underestimate the conditional CC of the high degree nodes.

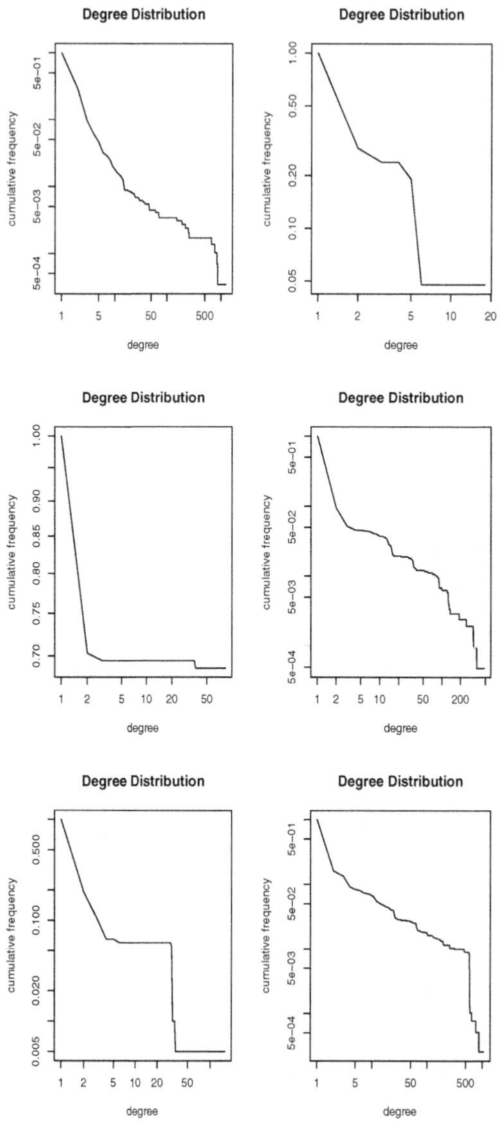

Fig. 3 Cumulative degree distributions of the eight ego networks

To measure how close our estimated distributions are to the real distribution, we need to define a distance (difference) measure that quantifies to what extent the estimated results are similar to the truth. Of course, the selection of a distance is crucial for the outcome of a study. Many distance measures have been defined in the literature. We use both Kolmogorov-Smirnov (KS) D statistic and Kullback-Leibler Divergence. The KS D statistic is based on the maximum distance between the two

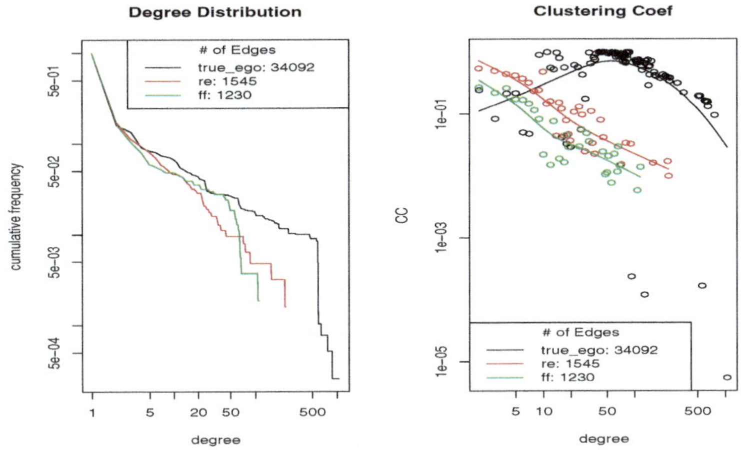

Fig. 4 Cumulative degree distribution and clustering coefficient estimation for one ego network using a sample size of 5% of the total network. RE: Random Edge sampling. FF: Forest Fire sampling

cumulative probability distributions, $D = sup_x|F(i) - G(i)|$, where $F(i)$ and $G(i)$ are the cumulative distribution functions. One of the limitations of the KS D statistic is that it is more sensitive near the center of the distribution than at the tails. It might be misleading to use this test to indicate whether two distributions are similar.

The Kullback-Leibler divergence (KLD), on the other hand, is a measure of dissimilarity between two probability distributions in information theory. It is the expected log-likelihood ratio, $KLD = \sum_i f(i) log \frac{f(i)}{g(i)}$ where $f(i)$ represents the estimated probability distribution and $g(i)$ represents the true probability distribution. We use the KLD to have a measure for the whole body of the distribution so that impact of mis-estimations at certain middle range degrees would not be as influential as that in the KS statistic.

The final estimates presented in the following section are the averages across 20 replicates of the same experiments.

5.2 Results

Figures 5 displays the KS D statistics of the degree distribution estimations. The eight plots present the eight ego networks. The number of nodes and the number of edges of the true ego networks are labeled as "V" and "E" respectively in the subtitles. Each line represents a different sampling method: solid lines represent the RE, and dotted lines represent our method (labeled FF as it is based on the Forest Fire algorithm). We can see that six out of eight the dotted lines are below the solid lines when sample size (x-axis) is bigger enough (20-50%). Furthermore, using the KS D measure, our method converges as the sample size approaches to

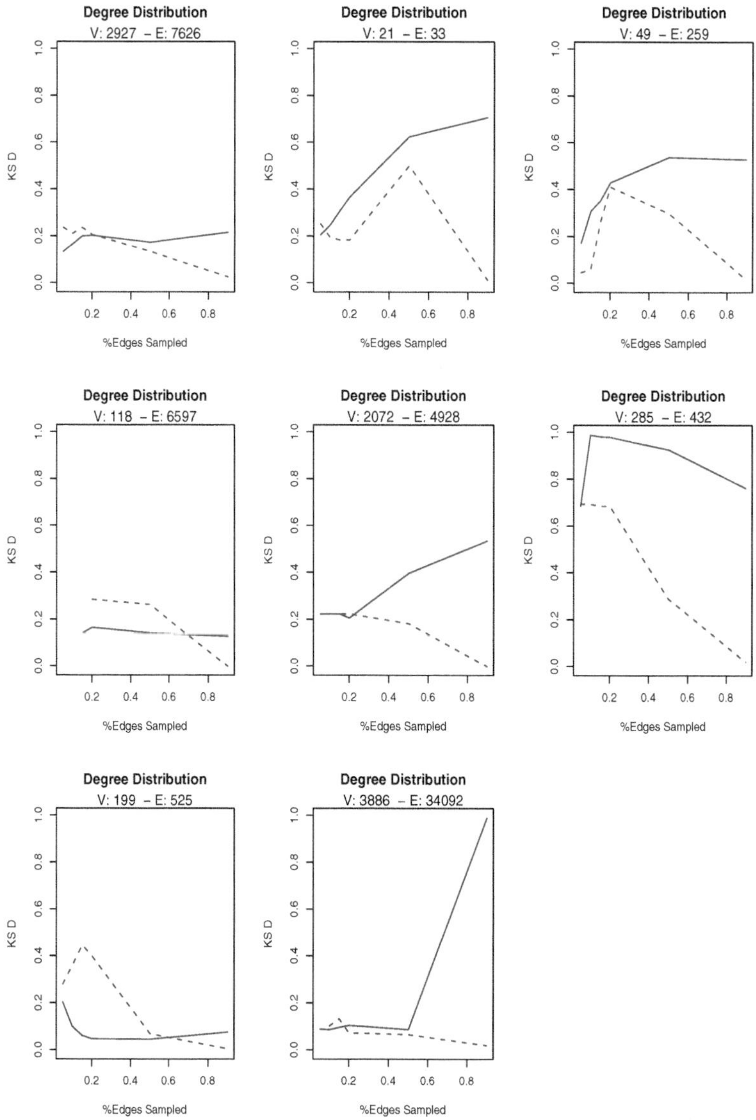

Fig. 5 Degree Distribution Estimation (KS D)

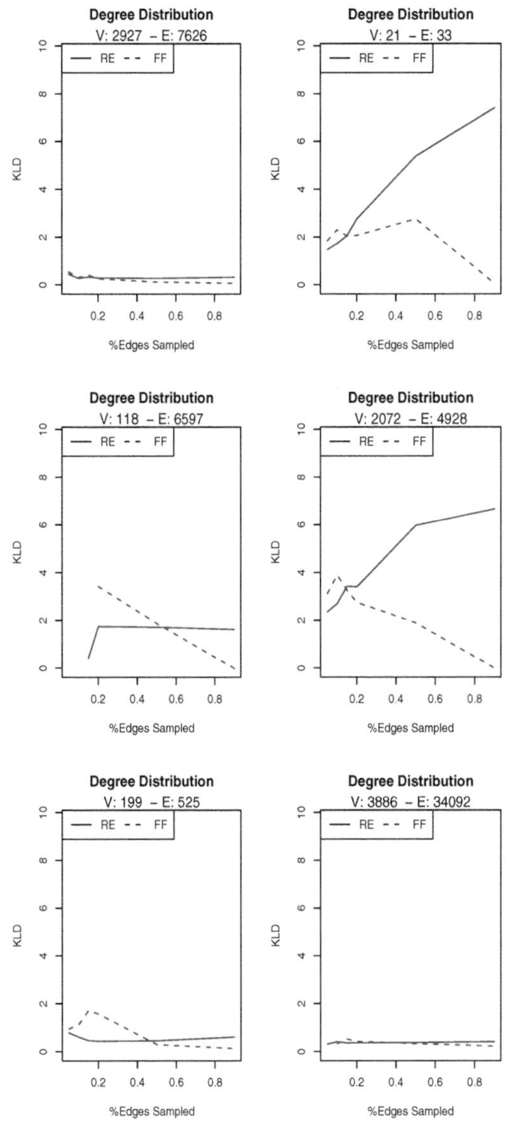

Fig. 6 Degree Distribution Estimation (KLD)

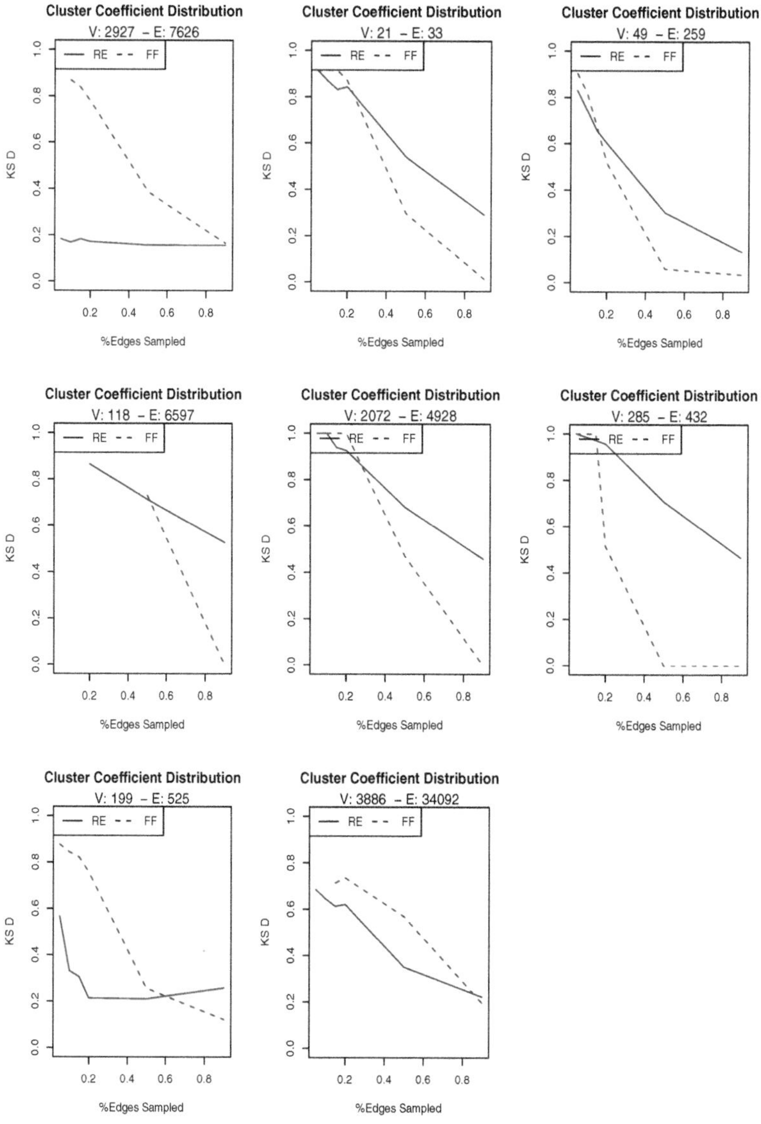

Fig. 7 Cluster Coefficient Distribution Estimation (KS D)

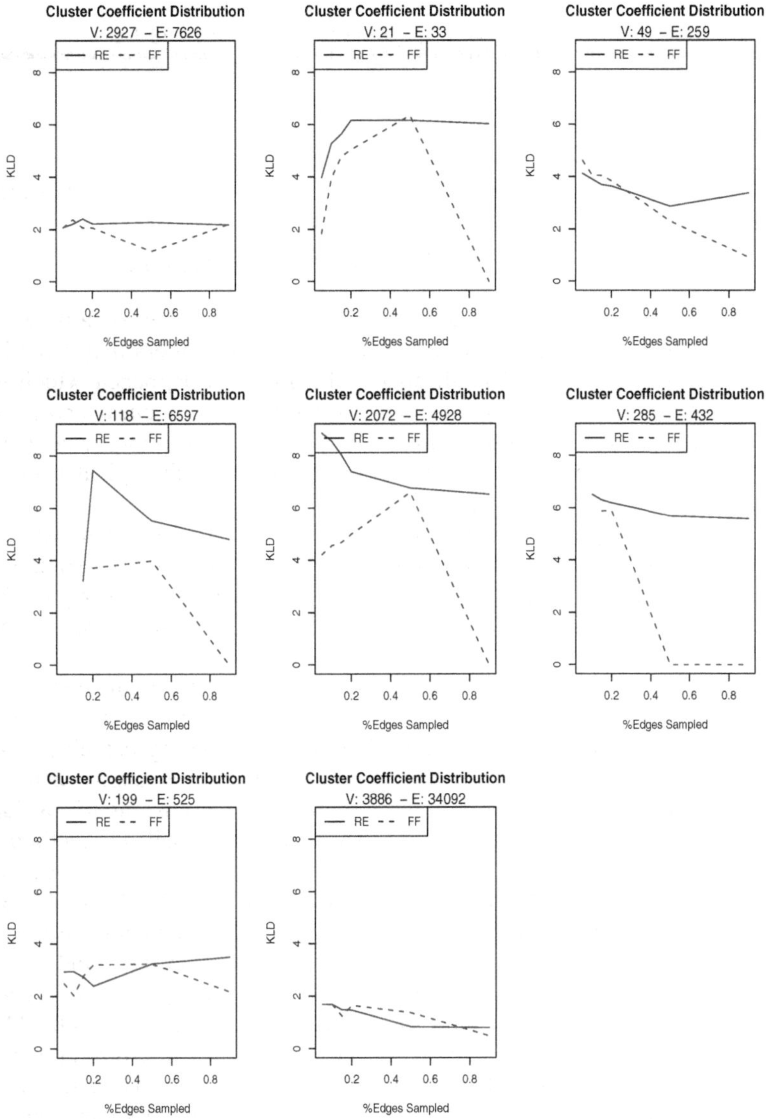

Fig. 8 Cluster Coefficient Distribution Estimation (KLD)

100%, while the RE method does not. Also, for more than half of the cases, the KS *D* statistic becomes worse as sample size increases when using RE. We believe there are fundamental flaws in this method. It fails to capture the dependence structure between nodes and edges.

Figure 6 summarizes the same results using the KLD measure. Again the FF method tends to converge toward 0 whether sample size increases, and is generally lower (better) than the RE method at all sample sizes.

Figure 7 displays the KS *D* statistics of the cluster coefficient distribution estimations. Unlikely the degree distribution, the the cluster coefficient distributions are not cumulative probability distributions. In order to apply the KS *D* statistic, they need to be transformed and standardized. Hence, the KS *D* statistic won't be able to detect systematic under- or over- estimations. Figure 8 displays the KLD statistics of the cluster coefficient distribution estimations. Based on these two figures. We can draw a similar conclusion: our approach outperforms the RE o method in accuracy and improvement in accuracy with larger sample sizes.

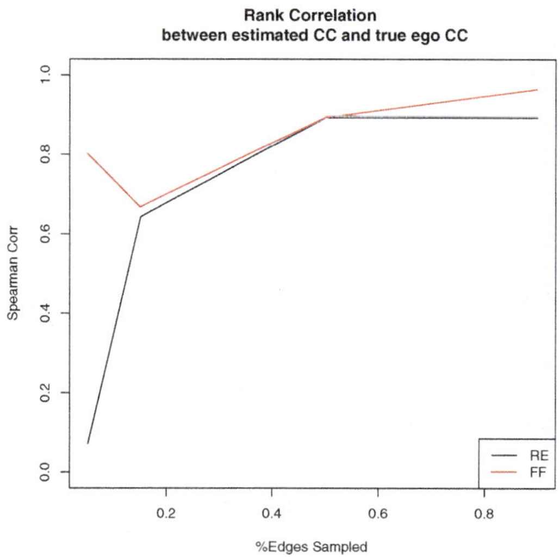

Fig. 9 Correlation between estimated CC and true CC

In contrast to Figures 8 and 7, Figure 9 looks at the unconditional CC of the ego network, i.e., the average CC of all nodes. It shows that the rank correlation (Spearman correlation) between the estimated CC and the true ego CC is very strong. We can see that even with a small sample size, e.g., $20-30\%$ of the total edges, we are able to estimate the rank of the ego networks by CC reasonably well. In other words, we can tell which web sites have richer local structure only by relatively small ego sampling. This is a significant validation of our ego-centric approach to viral marketing.

Figure 10 demonstrates that the ego networks can be used in estimating the global properties of the web sites (ego, or center nodes). It displays the rank correlation between the size of the ego network and the Page Rank[5] score [3] of the center node in the entire network. The ego network sizes of the web sites can be used as a proxy to estimate their relative Page Rank rankings. With small network samples, $< 20\%$, we can get decent Page Rank ranking estimations.

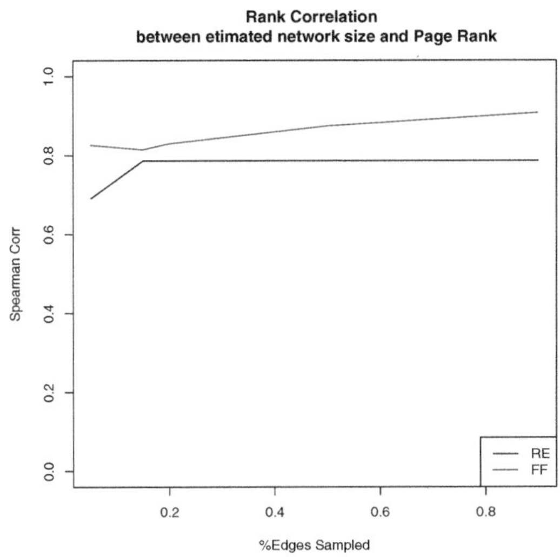

Fig. 10 Correlation between the size of the sampled ego-network and the Page Rank of the true ego-network

6 Conclusions

In this book chapter, we propose to use in-link networks to help determine the attributes of a web site and measure the effectiveness of viral marketing. We develop an ego-centric sampling approach for capturing the in-link network structure. We verify that the proposed sampling approach works well in capturing ego network structures, as long as the sample size is reasonable (20-30%). It outperforms the RE method in terms of estimating the underlying degree and cluster coefficient distributions of the ego networks, and converges asymptotically. We demonstrate that the ego-centric approach has interesting application potential in determining the global properties of the web sites and adds new information to in-links.

A major limitation of the approach lies in the fact that it is an ego-centric and sampling approach. It cannot be used to evaluate the existence of nodes/links and to find the shortest paths between web sites. Another limitation is that the sample in-link network does not give the full picture of the entire ego network, since

[5] We use the ARPACK implementation of Page Rank found in the R igraph package.

the out-links are collected passively, i.e., not all out-links can be collected, even asymptotically.

Much work still remains. The accuracy of the approach can be enhanced by sequential sampling methods: in case the collected data may not be representative of the ego network, we can increase the sample size in a sequential way until the measurements of certain statistics reach a steady state. Furthermore, investigations on the measurement bias as a function of sample size would be valuable.

References

1. Bar-Yossef, Z., Mashiach, L.T.: Local approximation of pagerank and reverse pagerank. In: CIKM 2008: Proceeding of the 17th ACM conference on Information and knowledge management, pp. 279–288. ACM, New York (2008), http://doi.acm.org/10.1145/1458082.1458122
2. Bharat, K., Chang, B.W., Henzinger, M.R., Ruhl, M.: Who links to whom: Mining linkage between web sites. In: Proceedings of the IEEE International Conference on Data Mining (2001)
3. Brin, S., Page, L.: The anatomy of a large-scale hypertextual web search engine. Computer Networks and ISDN Systems 30(1-7), 107–117 (1998), http://dx.doi.org/10.1016/S0169-75529800110-X
4. Domingos, P.: Mining social networks for viral marketing. IEEE Intelligent Systems 20(1), 80–82 (2005)
5. Faloutsos, J.L.C.: Sampling from large graphs. In: KDD 2006: Proceedings of the 12th ACM SIGKDD international conference on Knowledge discovery and data mining, pp. 631–636. ACM, New York (2006), http://doi.acm.org/10.1145/1150402.1150479
6. Henzinger, M.R., Heydon, A., Mitzenmacher, M., Najork, M.: On near-uniform url sampling. Comput. Netw. 33(1-6), 295–308 (2000), http://dx.doi.org/10.1016/S1389-12860000055-4
7. Kamvar, S., Haveliwala, T., Manning, C., Golub, G.: Exploiting the block structure of the web for computing pagerank. Tech. rep. Stanford University (2003)
8. Kleinberg, J.M.: Authoritative sources in a hyperlinked environment. J. ACM 46(5), 604–632 (1999), http://doi.acm.org/10.1145/324133.324140
9. Latapy, M., Magnien, C.: Complex network measurements: Estimating the relevance of observed properties, pp. 1660–1668 (2008), doi:10.1109/INFOCOM.2008.227
10. Mislove, A., Marcon, M., Gummadi, K.P., Druschel, P., Bhattacharjee, B.: Measurement and analysis of online social networks. In: IMC 2007: Proceedings of the 7th ACM SIGCOMM conference on Internet measurement, pp. 29–42. ACM, New York (2007), http://doi.acm.org/10.1145/1298306.1298311
11. Newman, M., Watts, D., Strogatz, S.: Random graph models of social networks. Proc. Natl. Acad. Sci. (to appear) (2002)
12. Newman, M.E.J.: Ego-centered networks and the ripple effect. Social Networks 25(1), 83–95 (2003), doi:10.1016/S0378-8733(02)00039-4
13. Park, H.W.: Hyperlink network analysis: A new method for the study of social structure on the web. Connections 25(1), 49–61 (2003)
14. Reicheld, F.: The one number you need to grow. Harvard Business Review 81, 47–54 (2003)

15. Richardson, M., Domingos, P.: Mining knowledge-sharing sites for viral marketing. In: Proceedings of the Eighth ACM SIGKDD International Conference on Knowledge Discovery and Data Mining, pp. 61–70. ACM, New York (2002), http://doi.acm.org/10.1145/775047.775057
16. Salganik, M.J., Heckathorn, D.D.: Sampling and estimation in hidden populations using respondent-driven sampling. Sociological Methodology 34(1), 193–240 (2004)
17. Yook, S.-H., Jeong, H., Barabási, A.-L.: Modeling the internet's large-scale topology. PNAS 99, 13,382–13,386 (2002)
18. Valente, T.W., Coronges, K., Lakon, C., Costenbader, E.: How correlated are network centrality measures? Connections 28(1), 16–26 (2008)
19. Valente, T.W., Davis, R.L.: Accelerating the diffusion of innovations using opinion leaders. The Annals of the American Academy of the Political and Social Sciences 566, 55–67 (1999)
20. Wasserman, S., Faust, K.: Social Network Analysis: Methods and Applications. Cambridge University Press, Cambridge (1994)
21. Watts, D., Dodds, P.: Networks, influence, and public opinion formation. Journal of Consumer Research 34(4), 441–458 (2007), http://research.yahoo.com/files/w_d_JCR.pdf
22. Wellman, B., Carrington, P., Hall, A.: Social Structures: A Network Analysis. In: Networks as Personal Communities, pp. 130–184. Cambridge University Press, Cambridge (1988)

Integrating SNA and DM Technology into HR Practice and Research: Layoff Prediction Model

Hui-Ju Wu, I-Hsien Ting, and Huo-Tsan Chang

Abstract. Recent developments in social network analysis (SNA) and data mining (DM) technology have opened up new frontiers for human resource management (HRM). SNA appears to be an effective tool for mapping relationships in an organization. The increased use of information technology provides useful new data about the user behavior automatically stored in database or web log files. Data mining methods were applied in practice to explore information from this huge amount of data. Data mining can be used to gain insight into the usage behavior based on objective data in contrast to subjective data. In this chapter we suggest ways in which combine SNA and DM be analyzed using network software and DM tool. We propose an example used exploratory research design conducting a single case study in Taiwan. This research aims at introducing the importance of the application of DM and SNA to predict layoff through an empirical study.

1 Introduction

The rapid development of information technology has also boosted the implementation of Human Resource Management. Information technology is driver to some present and upcoming changes in HRM. Data mining methods were applied in practice to explore information from this huge amount of data. Data mining can be used to gain insight into the usage behavior based on objective data in contrast to subjective data.SNA appears to be an effective tool for mapping relationships in an organization. We propose an example used exploratory research design conducting a single case study in Taiwan. This research aims at introducing the

Hui-Ju Wu · Huo-Tsan Chang
Graduate Institute of Human Resource Management,
National Changhua University of Education, Taiwan
e-mail: d94311001@mail.ncue.edu.tw, changht@cc.ncue.edu.tw

I-Hsien Ting
Department of Information Management,
National University of Kaohsiung, Taiwan
e-mail: iting@nuk.edu.tw

I.-H. Ting et al. (Eds.): Mining and Analyzing Social Networks, SCI 288, pp. 53–66.
springerlink.com

importance of the application of DM and SNA to predict layoff through an empirical study.

Global economic recession has been causing the unpaid leave and massive layoffs in major high-tech firms of Taiwan, both factors present great potential hardship to many employees according to the reports from industry. Therefore, layoff prediction and management have become great concerns of employees and managers. Employees wish to retain their jobs and keep their work for a long time. Hence, they need to predict the possible layoff and then utilize their resources to retain their job. In response to the difficulty of layoff prediction, this study applies social networks and data mining techniques to build a model for layoff prediction.

Previous researches on employees' turnover behavior mainly focus on the reasoning and affecting for employees' turnover intention. However, the factors for layoff and the construction of layoff prediction model from real business data still have not been well examined. Moreover, the application of social network analysis with data mining techniques for layoff prediction model construction is less addressed as well.

Therefore, layoff prediction and management have become of great concern to the employees and managers. Employees wish to retain their jobs and keep their work for a long time. Hence, they need to predict the possible layoff and then utilize their resources to retain their job. In response to the difficulty of layoff prediction, this study applies SNA and DM techniques to build a model for layoff prediction. Social network analysis treats organizations in society as a system of objects (e.g. people, groups, and organizations) joined by variety of relationships [11]. A research on social networks indicates that network structure and activities influence employees and affect individual organizational outcomes [13]. Data mining is thus emerging as a class of analytical techniques that goes beyond statistics and aims at examining large quantities of data in database.

This chapter aims at introducing the importance of the application of DM and SNA to predict layoff through an empirical study. It first provides a literature review on the recent research and application of SNA and DM. It is followed by a discussion of the concepts of DM and SNA. A case study based on an organization is then used to illustrate how SNA and DM is applied to develop a model for predicting the layoff. Future directions in applying SNA and DM in the organization networks have also provided in this paper.

2 Literature Review

2.1 Social Network

Social network analysis provides a rich and systematic means of assessing such network by mapping and analyzing relationships among people, teams, departments or even the entire organization [10]. Organizations are considered as a network of individuals and researchers have used network analysis to map information flow as well as relational characteristics among strategically important groups to improve knowledge creation and sharing [5]. Mapping and understanding social networks within an organization is a mean for us to understand how social relationships may affect business processes. To understand the complexity of the task, let us consider the various structural measures that can be applied to social networks. These structures are characterized by relationships, entities, context, configurations, and temporal stability. Some of the indices and dimensions that express outcomes of network are:

1. Size: density and degree. Size is critical for the structure of social relations due to each actor has limit resources and for building and maintaining ties. The degree of an actor is defined as the sum of the connections between the actor and others. The density measurement can be used to analyze the connectivity and the degree of nodes and links in a social network [14].
2. Centrality: The centrality of a social network is a measurement that is used to measure the betweenness and closeness of the social network. The measure of centrality which can be used to identify who have the most connections to others in the network (high degree) or whose departure would cause the network to fall apart [14].
3. Structural hole: The structural hole is also a measurement of social network analysis, which can be used to discover the holes in a social network and by this to fill the hole and expand the social network [14].
4. Reachability: The reachability can be used to analyze how to reach a node from another node in the social networks. An actor is reachable by another if there exists any set of connections by which we can trace from the source to the target actor, regardless of how many others fall between them [7].
5. Distance. Because most individuals are not usually connected directly to most other individuals in a population, it can be quite important to go beyond simply examining the immediate connections of actors, and the overall density of direct connections in populations. Walk, trail and path are basic concepts to develop more powerful ways of describing various aspects of the distances among actors in a network[4] [12].

2.2 Data Mining

Data mining has given the cleaned data intelligent methods that can be applied in order to extract data patterns. Data Mining is the extraction of hidden predictive information from large databases, is a powerful new technology with great potential to help companies focus on the most important information in their huge database [15].

Data mining technologies can be use to generate new business opportunities by providing capabilities if given databases of sufficient size and quality: automatic prediction of trends and behaviors, and automatic discovery of previously unknown patterns. The mostly common used techniques in data mining are listed as followings [15]:

1. Classification: The goal of classification is to predict the value of a user-specified goal attribute based on the values of other attributes, called the predicting attributes. This is the most studied data mining approach [6] [8].
2. Clustering: In clustering applications, data mining algorithms must "discover" classes by partitioning the whole data set into several clusters, which is a form of unsupervised learning[2].
3. Associations: it is unique feature with the capability to find association rules for items in a transaction file, and the capability to find all rules including compound and hierarchical rule [2].
4. Genetic Algorithm: Optimization techniques that use process such as genetic combination, mutation, and natural selection in a design based on the concepts of evolution [3].
5. Decision Tree: Decision tree technique is one of the data mining methods developed for classification and prediction. It is still of great help to reveal explicit relationships between attributes among huge data. Many researchers have been done with decision tree algorithm because of the great rule extraction and prediction ability [15].

3 Methods

3.1 Research Process

After clarifying the research background and objectives, we must now define the process and architecture of the research. To achieve the research goal, left employees' database of a semiconductor company at the Hsinchu Science Park of Taiwan will be used. The research architecture is shown in Figure 1, and the descriptions of each stage are presented as below:

1. **Step 1: Exploring Data** Analysis. The first stage to discover the record files of layoff from the left employees' databases.
2. **Step 2: Constructing Organizational Network.** Using social networks analysis for constructing a organizational network from the left employees databases. We use some network indicators, including density, degree, reachability, centrality, position and role to analyze the relationship between the manager and the laid-off employees.
3. **Step 3: Data Mining Analysis for the Layoff's file.** Applying data mining techniques for analysis the employees' attributes. We used the cluster analysis to discover classes by partitioning the whole data set into several clusters and used the association rules to discover the important associations among items.
4. **Step 4: Constructing Layoff Prediction Model.** Finally, we used the decision tree technique for classification and construction the layoff prediction model from the laid-off employees' organizational networks.

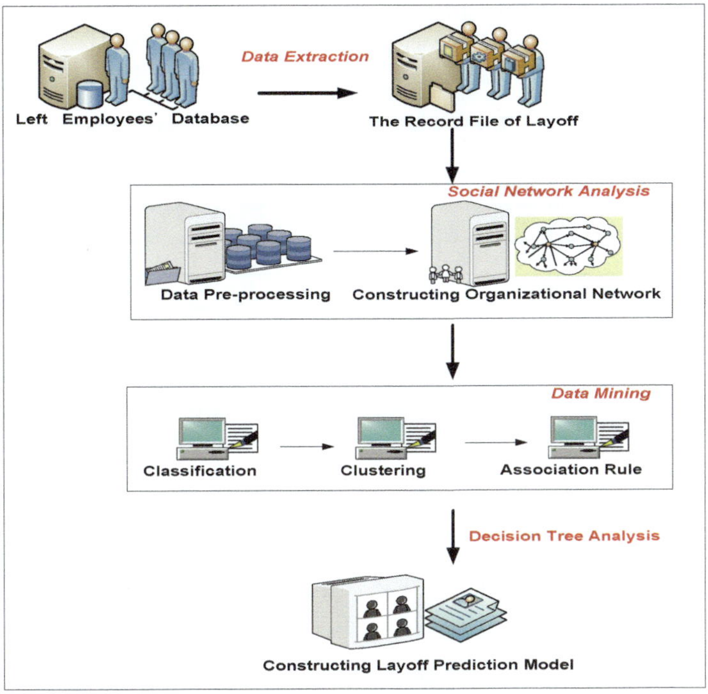

Fig. 1 Research Architecture

3.2 Constructing Organizational Network

The global economic recession has been causing the unpaid leave and massive layoffs by major high-tech firms in Taiwan. To explore the phenomenon, we used 528 left job employees records during 2007~2009 to test the empirical study. We propose ten attributes to build a organizational network by using the social networks analysis. These attributes include department, supervisor, sex, age, shift, live register, marriage, position, education level and grade.

Social network analysis is appropriate for "relational data". We need to construct the similarity attributes between managers and laid-off employees based on the social network of layoff. According the data with similar attribute values, tend to be assigned to the same networks and exist a relationship, whereas data different from each other tend to be assigned to distinct networks. In this network, The similarity attributes of employee and employee(manager) are than three kinds. We can indicate that has a tie between employee and manager. The relationship matrix is shown in Table 1. A1 has a tie with employees A3 and A6, but not with A2, A4, A5, SU_A m_A(Manager A) and not with him/her self.

Using social networks analysis for construct a organizational networks relationship of laid-off employees from the left employees databases. The process is described as Figure 2.This study proposed the example of organizational network A to explanation the research process.

Table 1 The relationship matrix of laid-off employees

	em_{A1}	em_{A2}	em_{A3}	em_{A4}	em_{A5}	em_{A6}	su_A
em_{A1}	0	0	1	0	0	1	0
em_{A2}	0	0	0	0	1	0	0
em_{A3}	1	0	0	0	1	0	0
em_{A4}	0	0	0	0	0	1	0
em_{A5}	0	1	1	0	0	1	1
em_{A6}	1	0	0	1	1	0	1
su_A	0	0	0	0	1	1	0

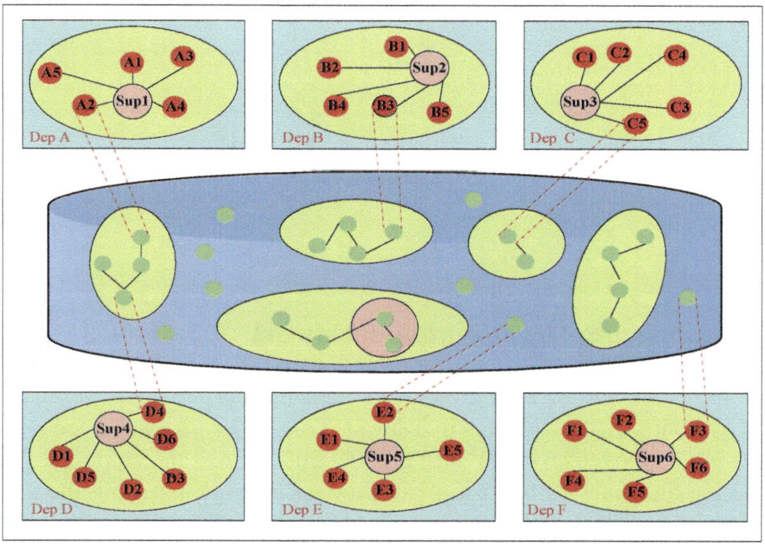

Fig. 2 Organizational Networks of laid-off employees

The study uses network software UCINET 6.182 to analyze the laid-off employees organizational networks indicators, including size, reachability, centrality, distance, and position and role analysis. Due to different stress on network, those indicators separately give us insight on how and for what degree they communicate with each other. The study is capable of finding some clues about network position for layoff. Then these variables will be used to construct network structure graph and data. With the approach, the research hopes to presume that the structure or pattern of ties in a social network is meaningful to the members of the network. The descriptions of each indicator as followed:

1. Size: density and degree

The density of a network is an examination how many correlation there are between employees compared to the maximum possible number of connections that exist between employees. The following Figure 3 is the density of this network. The density of network is 0.3816 based on the 16 ties.

Density

No.	Density	of Ties
Dep A	0.3810	16.0000

Fig. 3 Density of the Network

The following Figure 4 is the descriptive statistics of this network. The network degree centrality is 40% which describes this network centralization.

Degree

	1 Degree	2 NrmDegree	3 Share
5 emA5	4.000	66.667	0.250
6 emA6	4.000	66.667	0.250
1 emA1	2.000	33.333	0.125
7 suA	2.000	33.333	0.125
3 emA3	2.000	33.333	0.125
2 emA2	1.000	16.667	0.063
4 emA4	1.000	16.667	0.063

DESCRIPTIVE STATISTICS

	1 Degree	2 NrmDegree	3 Share
1 Mean	2.286	38.095	0.143
2 Std Dev	1.161	19.343	0.073
3 Sum	16.000	266.667	1.000
4 Variance	1.347	374.150	0.005
5 SSQ	46.000	12777.776	0.180
6 MCSSQ	9.429	2619.047	0.037
7 Euc Norm	6.782	113.039	0.424
8 Minimum	1.000	16.667	0.063
9 Maximum	4.000	66.667	0.250

Network Centralization = 40.00%
Heterogeneity = 17.97%.
Normalized = 4.30%

Fig. 4 Degree of the network

2. Reachability

The reachability can be used to analyze how to reach a node from another node in the social networks. The following Figure 5 is for each pair of nodes, the algorithm finds whether there exists a path of any length that connects them.

Reachability

	1	2 e	3 e	4 e	5 e	6 e	7 s
1 emA1	0	1	1	1	1	1	1
2 emA2	1	0	1	1	1	1	1
3 emA3	1	1	0	1	1	1	1
4 emA4	1	1	1	0	1	1	1
5 emA5	1	1	1	1	0	1	1
6 emA6	1	1	1	1	1	0	1
7 suA	1	1	1	1	1	1	0

Fig. 5 Reachability of the network

3. Centrality

The closeness centrality of a vertex is relied on the distance between one vertex and other vertices, which means that larger distances yield lower closeness centrality scores [12]. The following Figure 6 is the descriptive statistics of this network. The network closeness is 41.65% which describes this network centralization.

Closeness Centrality Measures

Statistics

	1 Farness	2 nCloseness
1 Mean	10.571	58.771
2 Std Dev	1.917	11.227
3 Sum	74.000	411.399
4 Variance	3.673	126.052
5 SSQ	808.000	25060.770
6 MCSSQ	25.714	882.367
7 Euc Norm	28.425	158.306
8 Minimum	8.000	46.154
9 Maximum	13.000	75.000

Network Centralization = 41.65%

Fig. 6 Closeness centrality

The betweenness centrality of an actor is the portion of whole geodesics between pairs of other vertices that contain this vertex. The following Figure 6 is the descriptive statistics of this network. The group betweenness is 36.67% which describes this network centralization.

Betweenness Centrality Measures

Un-normalized centralization: 33.000

	1 Betweenness	2 nBetweenness
5 emA5	7.000	46.667
6 emA6	7.000	46.667
1 emA1	1.000	6.667
3 emA3	1.000	6.667
4 emA4	0.000	0.000
2 emA2	0.000	0.000
7 suA	0.000	0.000

DESCRIPTIVE STATISTICS

	1 Betweenness	2 nBetweenness
1 Mean	2.286	15.238
2 Std Dev	3.010	20.068
3 Sum	16.000	106.667
4 Variance	9.061	402.721
5 SSQ	100.000	4444.445
6 MCSSQ	63.429	2819.048
7 Euc Norm	10.000	66.667
8 Minimum	0.000	0.000
9 Maximum	7.000	46.667

Network Centralization Index = 36.67%

Fig. 7 Betweenness centrality

4. Distance

The following Figure 8 can be quite important to go beyond simply examining the immediate connections of employees, and the overall density of direct connections in populations. Walk, trail and path are basic concepts to develop more powerful ways of describing various aspects of the distances among employees in a network.

Distance

For each pair of nodes, the algorithm finds the # of edges in the shortest path between them.
Average distance = 1.762
Distance-based cohesion ("Compactness") = 0.667
Distance-weighted fragmentation ("Breadth") = 0.333

Geodesic Distances

```
        1 2 3 4 5 6 7
        e e e e e e s
        - - - - - - -
1 emA1  0 3 1 2 2 1 2
2 emA2  3 0 2 3 1 2 2
3 emA3  1 2 0 3 1 2 2
4 emA4  2 3 3 0 2 1 2
5 emA5  2 1 1 2 0 1 1
6 emA6  1 2 2 1 1 0 1
7 suA   2 2 2 2 1 1 0
```

Fig. 8 Distance analysis

The following Figure 9 is a set of points and a set of lines between pairs of points. Points and lines understood in graph display employees and their ties known in social network analysis, directed graphs with one or two way arrows are used to display the degree of correlation between employees.

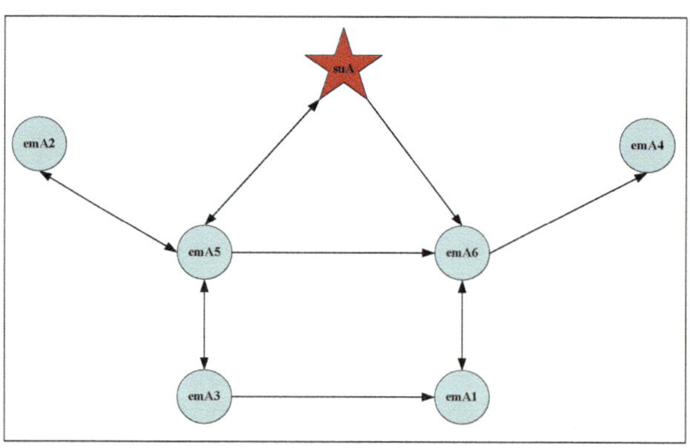

Fig. 9 Graph analysis

5. Position and role Analysis

The position and role Analysis define the social position as collections of employees who are similar in their tie with others and modeling social roles as systems of ties between employees or between position.

Position and role

Structural Equivalence Matrix

	emA1	emA2	emA3	emA4	emA5	emA6	suA
	-----	-----	-----	-----	-----	-----	-----
emA1	0.00	2.45	2.00	1.41	2.00	2.83	2.00
emA2	2.45	0.00	1.41	2.00	2.45	2.45	1.41
emA3	2.00	1.41	0.00	2.45	2.83	2.00	2.00
emA4	1.41	2.00	2.45	0.00	2.45	2.45	1.41
emA5	2.00	2.45	2.83	2.45	0.00	2.83	2.00
emA6	2.83	2.45	2.00	2.45	2.83	0.00	2.00
suA	2.00	1.41	2.00	1.41	2.00	2.00	0.00

Fig. 10 Position and role Analysis

The following Figure 11 is dendrogram for complete link hierarchical clustering of Euclidean distances on the relation for employees. To compare Euclidean distance, the short distance is 1.414 between the employee A2 and A3 d. The employee A2 and A3 have the similar position and classify to the same cluster.

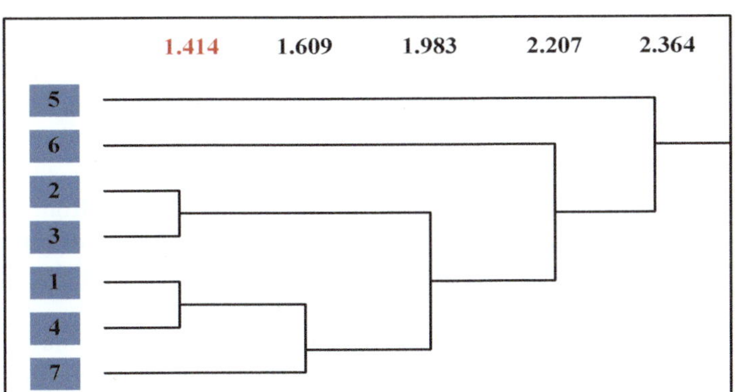

Fig. 11 Clustering analysis of the position

3.3 Data Mining Analysis for Layoff's File

In this session, in order to construct the layoff prediction model, we used the data mining techniques for extracting rules from selected data. This research used 124 training data from the laid-off employees network of a semiconductor company in the Hsinchu Taiwan Science Park. The testing data is 100 employees' data of the active employees' database from the same resource in the year 2009. Each record in employees' database consists of 15attributes. The original attributes of each column are as follows.

Table 2 The Attributes List

Attribute	Attribute	Attribute
1.ID	6. Compensation _LV	11. Hire_DT
2. Name	7. Live register	12.Termination_DT
3. Dept_ID	8. Education_LV	13.Supervisor_ID
4. Sex	9. Marriage	14. Position
5.Age	10. Grade	15. Shift_DESCR

The data mining techniques involved in this research are demonstrated, including feature selection techniques for diminishing the data dimension. The classification analysis and association rule for extracting rules from selected data. The descriptions of each analysis as followed:

1. **Clustering:** Clustering is the task of segmenting a heterogeneous population into a number of more homogeneous subgroups or clusters. According to the attributes, we selected age, sex, marriage, grade, education level, shift, position and compensation level to cluster the left employee 6 segments by K-Means. To keep all clusters almost the same number of employees, we firstly divided each variable into four parts by quantification. We then transformed these numeric data to be categorical one for clustering.
2. **Association Rule:** Association analysis is the discovery of association rules showing attribute value conditions that occur frequently together in a given set of data. The layoff of association rules are generated from WEKA tool that was developed by University of Waikato in New Zealand. We used WEKA to do association and found the 8 useful rules for the laid-off employees' attribute.
3. **Decision Tree:** A decision tree divide the records in the training set into disjoint subsets, each of which is described by a simple rule on one or more fields [3]. In this research, the training data set contains 124 records and the testing dataset has 100 records. This research combines the training dataset and testing dataset into a table.

We summarized research results that: 1. The decision-tree algorithm for fatty liver screening has an accuracy of 86.2%, and it is better than logistic regression; 2. The accuracy of decision-tree algorithm for moderate to severe fatty liver disease is 93%; 3. The cut points of six parameters in decision-tree algorithm are: 1. Layoff, Age=40~50, Sex=M, Mar=Y, Shi=Normal shift, Pos=Manager_Lv and Com ≧ 50000 had an approximately 86.2% accuracy rate for predicting the laid-off employees. 2. Layoff, Age=40~50, Sex=M, Mar=Y, Edu=University, Pos=Manager_Lv and Com≧50000 had an approximately 92.17% accuracy rate for predicting the laid-off employees. 3. Layoff, Age=40~50, Sex=M, Mar=Y, Gra≧10 , Pos=Manager_Lv and Com≧50000 had an approximately 96.44% accuracy rate for predicting the laid-off employees. In consequence, we find the high compensation, high position, high grade, high education level to become the dangerous list of layoff.

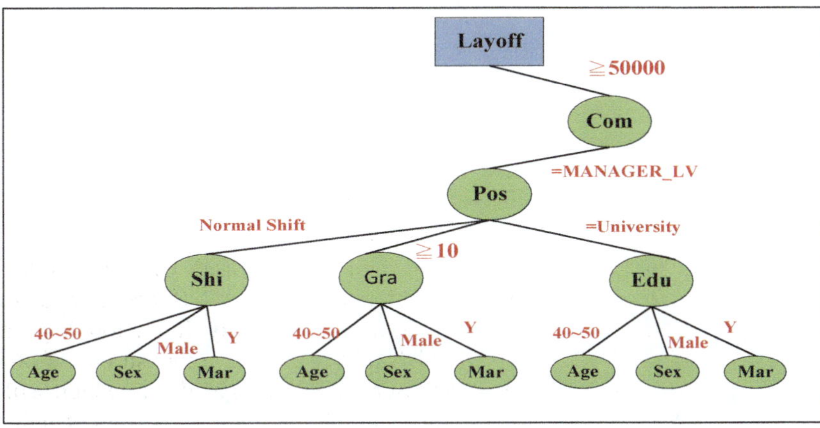

Fig. 12 Decision Tree analysis

4 Discussion and Conclusion

This chapter provided a new research direction of combing SNA and DM methods in HRM. We examine structural positions of individuals, especially HR actors (line managers and employees) within relational networks for building layoff prediction model and to explore implications for designing and implementing HR practices.

This study aims to verify the main causes for layoff factors. This research intends to base on these factors and concepts that addressed by above to find a best layoff predictive model using the social network and data mining techniques. Through an empirical evaluation, the results indicated that the proposed approach has pretty good prediction accuracy by using organizational networks relationship, employees databases, and layoff records to build layoff prediction model. Both decision tree techniques are good candidates to be applied to develop the model. The main aim of this study is to highlight how to predict layoff for employees and reduce the unemployed rates based on mining historical databases, and hopefully provide a layoff predictive model for employees and company.

Facing the global recession, within the Hi-Tech industry such as the semiconductor one of the challenges is to understand and retain the beneficial employees for company. The current trend of layoff cut many high compensation managers in Hi-Tech industry. It is important phenomenon to make one deep in thought for employees. This research data only forms a single semiconductor company in the Hsinchu Taiwan Science Park. The future research will apply this model to other industry.

References

[1] Burt, R.S.: The Network Structure of Social Capital. Research in Organizational Behavior 22, 345–423 (2000)

[2] Berson, A., Smith, S., Thearling, K.: Building Data Mining Applications for CRM. McGraw-Hill, New York (2000)

[3] Berry, M.J.A., Linoff, G.: Data mining techniques: For marketing, sales, and customer support. Wiley, New York (1997)

[4] Carrington, P.J., Scott, J., Wasserman, S. (eds.): Models and Methods in Social Network Analysis. Cambridge University Press, New York (2005)

[5] Cross, R., Parker, A.: The Hidden Power of Social Networks. Harvard University Press, Cambridge (2004)

[6] Han, J., Kamber, M.: Data Mining: Concepts and Techniques. Morgan Kaufmann Publishers, San Francisco (2001)

[7] Hanneman, R.A.: Introduction to Social Network Methods, University of California, Riverside (1998), `http://www.faculty.ucr.edu/~hanneman/`

[8] Kudo, M., Skalansky, J.: Comparison of Algorithms That Select Features for Pattern Classifiers. Pattern Recognition 33(1), 25–41 (2000)

[9] Kilduff, M., Wenpin, T.: Social Networks and Organizations. SAGE Publications, London (2003)

[10] Lutters, W.G., Ackerman, M.S., Boster, J., McDonald, D.W.: Creating a knowledge mapping instrument: approximation techniques for mapping knowledge networks in organizations. ICS Technical Report, No. 99–32), Center for Research on Information Technology and Organizations. University of California, Irvine, CA (2001)

[11] Škerlavaj, M., Dimovski, V.: Social Network Approach To Organizational Learning. Journal of Applied Business Research 22(2), 89–97 (2006)

[12] Nooy, W.D., Mrvar, A., Batagelj, V.: Exploratory Social Network Analysis with Pajek. Cambridge University Press, New York (2005)

[13] Sparrowe, R.T., Liden, R.C.: Wayne, & Kraimer,M.L. Social networks and the performance of individuals and groups. Academy of Management Journal 44(20), 316–325 (2001)

[14] Wasserman, S., Faust, K., Iacobucci, D., Granovetter, M.: Social network analysis: methods and applications (Currently unavailable) (1994)

[15] Kurt, T.: A Introduction of Data Mining. Direct Marketing Magazine (February 1999)

Actor Identification in Implicit Relational Data Sources

Michael Farrugia and Aaron Quigley

Abstract. Large scale network data sets have become increasingly accessible to researchers. While computer networks, networks of webpages and biological networks are all important sources of data, it is the study of social networks that is driving many new research questions. Researchers are finding that the popularity of online social networking sites may produce large dynamic data sets of actor connectivity. Sites such as Facebook have 250 million active users and LinkedIn 43 million active users. Such systems offer researchers potential access to rich large scale networks for study. However, while data sets can be collected directly from sources that specifically define the actors and ties between those actors, there are many other data sources that do not have an explicit network structure defined. To transform such non-relational data into a relational format two facets must be identified - the actors and the ties between the actors. In this chapter we survey a range of techniques that can be employed to identify unique actors when inferring networks from non explicit network data sets. We present our methods for unique node identification of social network actors in a business scenario where a unique node identifier is not available. We validate these methods through the study of a large scale real world case study of over 9 million records.

1 Introduction

Until quite recently the term "network" conjured up images of routers, cables and gateways in a computer scientist's mind. Today however, due to the explosion of social media web sites [6], the same word evokes images of "one's circle of friends". The same word, network, is used in different contexts, yet it is still referring to the same structure. A network is the name given to a structure that connects objects together. In the case of the physical network the objects were computers whereas in

Michael Farrugia · Aaron Quigley
UCD Dublin, Dublin, Ireland
e-mail: mike.farrugia@gmail.com, aquigley@ucd.ie

I.-H. Ting et al. (Eds.): Mining and Analyzing Social Networks, SCI 288, pp. 67–89.
springerlink.com

the second case of social media, it is people. There are many other manifestations of interconnected structures in the world where connection patterns can be described with networks.

Examples of networks that exist in the real world include the world wide web [2], transport networks [9], cell biology networks [5], software networks [26] and linguistic networks [52]. Albert and Barabási provide further examples of networks and their properties in [1].

By contrast to inanimate networks, social networks define a subset of networks that involve human interaction and human actors. In social networking terms, a network is defined as a set of actors (objects or *nodes* in inanimate networks) and a set of ties (connections or *edges*) between those actors [51]. Typical examples of actors are persons, organisations or countries. Ties represent relationships between the actors. Often relationships link actors based on some kind of interaction, which explains why links are described by verbs including; married-to, son-of, sought-advice-from, plays-with, emailed-to, donated-money-to, participated-with and talks-to.

1.1 Implicit and Explicit Network Data Sources

The distinguishing feature of network data is that this data is *relational* as opposed to simply *attribute* based. Relational data is based upon connections between data elements, whereas attribute data is based on the properties of the data elements and each data element is independent. The analytical techniques used for the two types of data also vary. Relational data is usually analysed using Social Network Analysis (SNA) [55], while attribute data is analysed using variable analysis.

Relational data can be either *explicitly* collected or *implicitly* extracted from the raw data. Explicit network data sources unambiguously define both the actors and the relationships between those actors. These data sources are typically collected with the explicit intent of analysing the data using social network analysis techniques instead of variable based analysis. Most manual data collection methods in sociological studies collect data using self reporting surveys where the relationships and the nature of the relationship is specified clearly in the survey [39, 59, 17].

With implicit data sources the actors, or more often their relationships, are not explicitly defined. These data sets are not collected with the intention of studying the relationships between actors, but intended to be analysed independently using variable analysis. The Al Qaeda network [40] extracted by Krebs, is a clear example of an implicit network. This network was constructed from media reports following the attacks of September 11 and was generated from a number of data sources, including news articles were the relationships are implied, or sometimes even speculative. In theory, social networking web-sites [16] form explicit networks as the focus of the data is on the relationships. The relationships between friends are explicitly defined and subsequently confirmed by both actors. In reality however, the abused definition of the friendship relationship can limit the sociological significance of the underlying social network. Familial, social, workplace and even sexual relationships are folded into the simple category of "friend" which misses the rich context we know exists. Data cleanliness and reliability is further a major concern when such systems are used to support gaming where unknown people are added as friends, to benefit in online games that reward a high number of relationships. Relationship simplification, misuse or abuse of such systems can all result in network data with skewed properties which are not representative of the either the real-world or even active online social activities.

Although attribute based analytical techniques assume independence between individual rows of data, in fact attribute heavy data sets can contain relationships that may be beneficial to analyse. When analysing attribute data together with a relational perspective a new dimension is added to the analysis which can provide valuable insight. For example, customer analysis is a traditionally attribute heavy domain where most analysis is performed with traditional attribute based analysis methods [11]. In this domain however, it has been recognised that a customer rarely behaves independently and it is advantageous to consider a relational perspective of the customer [35, 27, 31] ie. a customer has friends, family, partner and colleagues who may also be customers.

The immediate advantage of inferring social networks from attribute based systems is manifested in the increased scale and time span of the data. When relying on manual data collection the size of the data collected is typically limited to at most a few hundred actors. When networks are inferred from large automatically created data sets, the number of nodes can easily span thousands or millions of actors. The number of time points in manually created data sets is also limited by the feasible number of times a survey can be conducted. When using automatically collected data that is timestamped, extracting large dynamic networks becomes possible.

An additional advantage that comes with automatic network extraction is the elimination of self report bias when actors respond to network surveys. The bias can be introduced by a lack in the ability to remember instances, one's personal understanding of the relationship terms in the survey, and the reliance on the good will of the participant to supply accurate results [29]. The derived benefit however comes at a cost, as while intentional bias may be eliminated, automatically collected large scale data can be dirty and require various degrees of sanitisation.

While manual data collection might under report, automatic data collection processes may over report on relations. Automatic data collection processes are designed to be comprehensive and catch all instances of any event that occur. These events might include ones that are not relevant for data extraction. For example, in the case of a phone call network, a person may call their mailbox to retrieve messages or call their own phone to locate it when lost. While these are both valid calls and are recorded in call log, they are not significant to an extracted social network.

1.2 Network Inference Approach

In order to be able to infer social networks two artifacts must be identified, the actors and the relationships between the actors. In this chapter we concentrate on

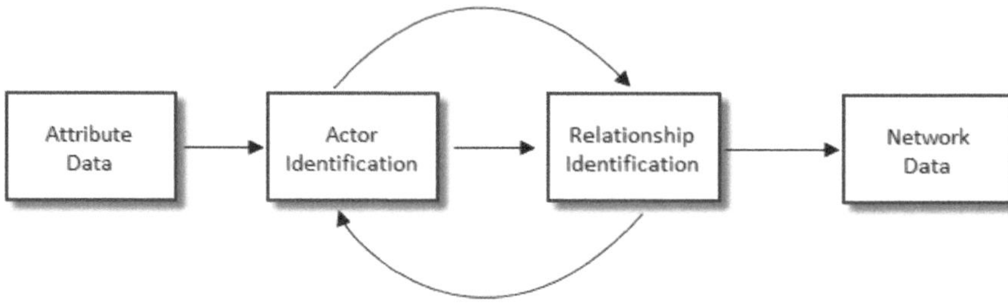

Fig. 1 Network Inference Process

the correct identification of network actors. This is a non-trivial process when the concept of "actor" isn't at the forefront of the system design which is collecting non-relational data. As such, this step involves identifying all the cases where actors are not properly represented, typically when appearing under different identifiers in different records. This task is critical if there is no unique identifier that identifies each actor unambiguously.

Network actor identification is a reformulation of the "entity resolution" problem that is frequently encountered in different areas of computer science (see section 3). Entity Resolution approaches can be divided into two categories, namely attribute based approaches and relational based approaches. Attribute based approaches, discussed in section 4.1, consider all the data elements independently and do not exploit relationships, whether present or not, between data elements. Relational approaches, discussed in section 4.2 use an identified network structure as additional information to improve the quality of the entity resolution.

When inferring a network, relationships are not always trivial to infer. Ambiguous definitions of relationships, different types of relationships, different measures of relationship strength [43, 46] and lack of concrete supporting evidence in the data, can make the process of relationship identification complex. Furthermore, if relationship data is not available then relational entity resolution techniques cannot be employed as there is no network data available.

In the network inference framework illustrated in Figure 1 we propose a cyclic process whereby actors are first resolved using attribute based entity resolution and then improved upon following the initial relationship identification stage. The cycle between identifying actors and identifying relationships can be refined progressively, in both directions. The relationship information can be used to improve the quality of matching identical entities, while the observations from actor identification can prompt rules to identify new types of relationships. In the second part of this chapter, in section 5, we use a real world case study to illustrate the steps within the first stage of actor identification using attribute information. Future work will describe how relationships are identified from the data set and how this information can be fed back to improve the quality of the actor identification process.

2 Rational for Identifying Unique Actors

If each actor in a non-relational dataset has a unique identifier then the process of actor identification is straightforward, apart from noisy or spurious entries. When, as is often the case, no unique identifier exists then the process is more challenging. Hence, the actor identification process typically involves matching records with similar personal information to the same person. Consider the 3 records with name, address and telephone fields shown in Table 1. After a cursory glance, based on the attributes given, one can infer that Thomas O'Connell and Tom Connell possibly refer to the same person, but the last Thomas Connell probably isn't the same person even though he has the same name as the person in record 1. Entity resolution is the technique used to automate this process.

Table 1 Similar example records for entity recognition

Name	Address 1	Address 2	Phone
Thomas O'Connell	15 Parnell Street	Dublin 2	085 123 4233
Tom Connell	Parnell Square, 15	Dublin 2	(0)85 123 4233
Thomas O'Connell	15 High Street	Dublin 1	+353 85 458 1112

In a social network each node of the network is a unique member of the network. If the same entity is present more than once in the network, then patterns and measures calculated from the social network will be inaccurate [15]. If the above example is extended to a social network, the importance of entity resolution is clear. Consider a social network derived from e-mail communication between a group of friends. Table 2 shows the original list of emails sent between the 6 friends, before entity recognition is applied.

Table 2 Email communication between friends before entity recognition

From	To
mary.jane	tom.connell
mary	james.home
michael.home	maria
mike.work	james.work

Figure 2 shows the network of the relationships before entity recognition has been applied. Notice that the network is fragmented, and is consists of 4 minimal components. If through entity recognition we identify that the e-mail names *michael.home* and *mike.work* are both referring to the same Michael, and *james.home* and *james.work* are referring to the same James, then the network now looks like the one shown in Figure 3.

Following the entity recognition stage the network changes drastically, reducing the number of components by half. Entity recognition is rarely 100% precise and

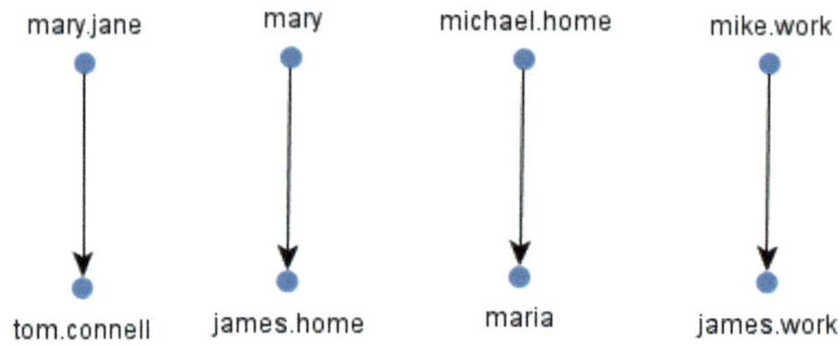

Fig. 2 Network of email communication before entity recognition

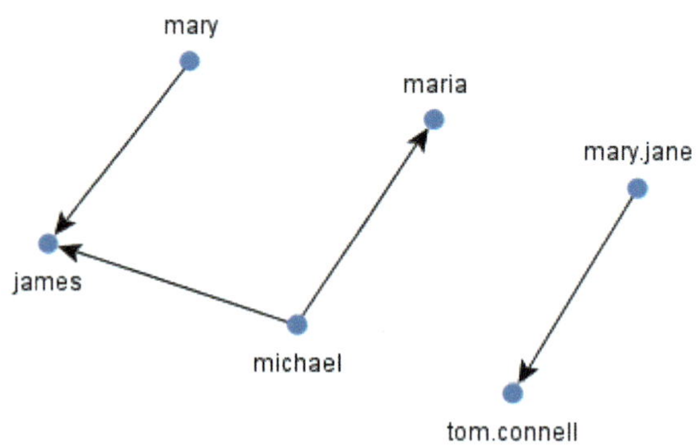

Fig. 3 Network of email communication after entity recognition

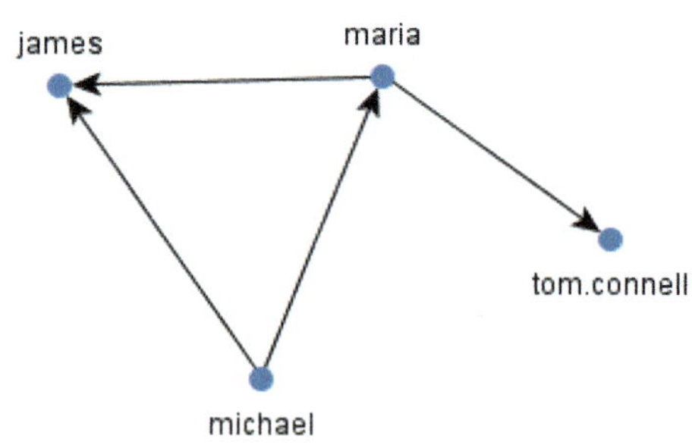

Fig. 4 Network of email communication after over matching entity recognition

"over matching" entities is a factor that can also adversely effect the interpretation of a network. In the case of "over matching" entities, there is a higher likelihood of hubs being identified when in reality they do not exists. Figure 4 shows the network if we incorrectly merge the records of *maria*, *mary* and *mary jane*.

Table 3 Effect of entity resolution on the social network

Measure	Under-matched	Correct	Overmatched
network density	low	medium	high
number of components	high	medium	low
network distance	high	medium	low

3 Entity Resolution

The problem of identifying multiple records referring to the same single entity was recognised more than six decades ago. The first definition of the problem came from H. L. Dunn [28] who used the term *record linkage* to define the problem. Later geneticist Howard Newcombe proposed some key approaches, including matching methods, that are still in use in today's systems [45]. The seminal paper by Fellegi and Sunter [32] from the statistics community formally defined record linkage building on prior work by Newcombe. Although the problem is well understood and has had considerable attention within the research and development community, it is still considered as one of data mining's grand challenges [47].

In computer science the same problem spans many different research communities, often under different names. In the database and KDD communities the problem is often called the *merge/purge*, *data cleansing* or *duplicate elimination* [34] problem. In this context, the aim is to identify which tuples within the same table or different tables, correspond to the same real world object. Computer scientists and AI practitioners refer to the problem as *entity resolution*. In computer vision the term *correspondence problem* [50] is used to describe the identification of features belonging to the same object in two different images. The problem has also been an open topic in Natural Language Processing, under the term *coreference resolution*. In NLP, coreference resolution is part of information extraction, where names referring to the same entity in free form text need to be identified as referring to the same person. The message understanding conferences (MUCs) sponsored by DARPA aided with the definition and evaluation of coreference resolution by introducing coreference tasks in the yearly challenge after its 6th conference [36].

The application of entity resolution has been applied and documented in several domains. The first applications were on medical data [45] and since then there have been more than a thousand references to articles on the subject published in medical literature [24]. Significant studies on US census data have been conducted by Winkler [58], and applied by national statistics bodies of other countries [53]. Entity resolution can also be used to identify fraud. For example, matching employment records with records to disability claims can uncover cases of disability compensation fraud [37]. Other examples include deduplicating lists of potential customer names for direct marketing [34] and deduplicating search results in meta search engines [13].

4 Entity Resolution Approaches

In entity resolution literature, perhaps due to the original work from the statistics community [58], the predominant approach is undertake a pairwise comparison between records. This involves comparing the data attributes to determine the similarity between pairs of records in order to classify the pair as a match or a non match. This approach is attribute based and considers each pair independently. Transitive closure is often calculated on the resulting pairs to merge records that are pointing to the same entity.

Since the problem has been tackled by different research communities, it has been formulated in a variety of ways [10], some of which exploit the nature of the data to infer more information than attributes alone can contain. Relational information such as child-parent relationships and co-authorship links between paper authors, can be used to create a graph of common neighbours which provides more information to make the entity resolution process more accurate.

4.1 Attribute Based Entity Recognition

A typical attribute based entity resolution solution is divided into five stages [22]. Before attempting to identify entities, the data available has to be cleaned and consistently divided into separate fields that are used in the following stages of entity resolution. After the cleaning stage, the data is typically divided into blocks to reduce the number of comparisons between potential duplicates. Next, field comparisons measure the similarity between pairs of records to enable a classification of which records are identical and which are not. Finally, the output of the classification is evaluated to measure the quality of the whole process. The following sections will describe each of these stages in more detail.

4.1.1 Data Cleansing and Data Standardisation

As the title of Hernandez's paper states, "Real word data is dirty" [34]. The entire process of entity recognition is itself often a preprocessing stage before data mining or analysis, which explains why entity resolution is sometimes referred to as *data cleansing* in the database community. Despite entity resolution being a cleansing stage for data analysis, the raw input data itself needs to be standardised in a single well defined common format prior to other stages in the entity resolution process.

Data cleansing is well a understood problem in the database and datawarehousing communities [49]. Leading database providers have commercialised research and are now providing tools specifically designed to assist in data cleansing as part of the ETL (extraction, transformation and loading) process to populate datawarehouses. At this stage of the entity resolution process the main concern is ill formatted data, different encodings of the same data, or data residing in incorrect fields.

Dates, addresses and phone numbers are typical examples of fields that require standardisation for entity recognition. For example in Table 1 the 3 phone numbers are encoded differently. In order to make comparison accurate this data must be

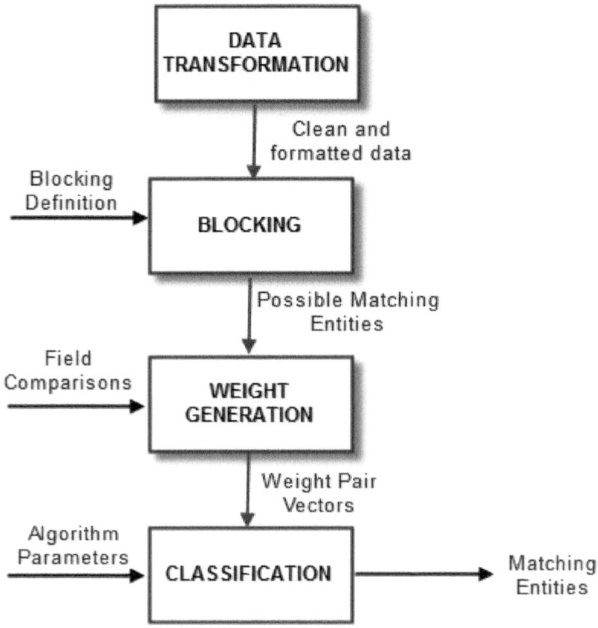

Fig. 5 The entity resolution process

standardised into a single common format. A fine grained division of data into separate fields, such as storing the street, city and country in separate fields, is important for high comparison accuracy. Approaches to automatically identify and standardise this data based on hidden Markov models (HMM) and traditional dictionary based lists have been studied by Churches et al. [24].

The data cleansing stage can have a direct effect on both the accuracy and speed of the entire resolution process. Although some comparison functions on strings can tolerate a threshold of dirty data, typically the more robust the function the more expensive it is in terms of execution time [44]. The individual field comparison method, is usually the most expensive aspect of any entity resolution process, therefore minimising the cost of these methods is desirable. The data cleansing stage can, for example, convert shortened names such as "Mike" into "Michael" using lists before the field comparison stage. Although string comparison functions have come a long way to identify similar stings, applying such data transformations in the data cleansing stage can help to improve accuracy. It is important to note that most of these transformations are domain dependent and specific to the type of data.

4.1.2 Blocking

If two data sets, A and B, are to be linked the complete number of comparisons is equal to the cross product of the size of the two total datasets, $|A| \times |B|$. When deduplicating a single data set the maximum number of comparisons is $\frac{|A|\times(|A|-1)}{2}$. For any commercial scale data set, the maximum number of comparisons rises rapidly. For example, if one were to link two data sets of 10,000 records each, the total number of comparisons is 100,000,000. The record comparison operation is the most expensive operation of the entire entity resolution process, therefore reducing the number of comparisons will improve the scaleability of the process.

Typically, the number of true matches from the total number of possible matches is usually only a very small fraction of the cross product of the two data sets. So if the two 10,000 record data sets were to overlap completely one to one, the number of matches is still only 10% of the total number of matches, and depending on the data sets this can be much less. A heuristic process called blocking [8] can be applied to reduce the number of comparisons, this partitions the dataset into different blocks that are likely to contain duplicate records. The records within these blocks are compared together, however records in one block are not compared with others from a different block. Figure 6 shows how segmenting a 1000 record dataset into 5 blocks of 200 records each, reduces the number of number of comparisons from one million to 200,000. Blocking methods have also been applied in many domains, in for example, the detection of repeated lines of program code in a large software system, so called "Clone Detection" [7].

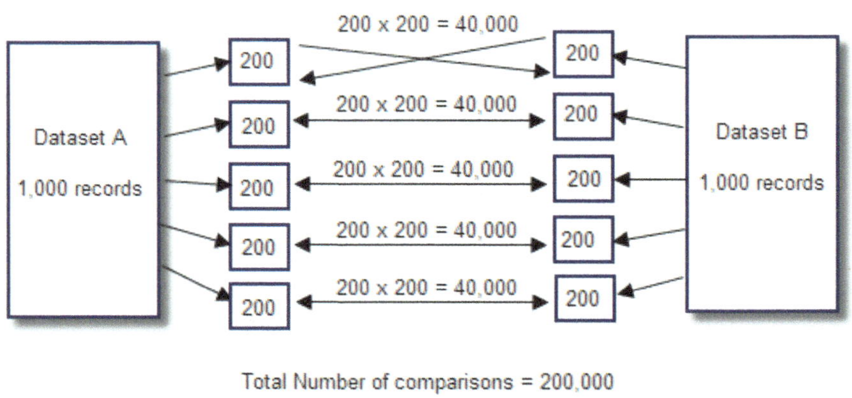

Fig. 6 Blocking of 1000 records (adapted from [53])

In order to partition the data a key is built from one or more fields of the data set in use. A typical example of a key, in data containing addresses, is the postcode. In this case only records belonging to the same post code are compared. As a result, the number of blocks is equal to the number of unique postcodes. A key of the length of a line of code is used in [7] to block a million line code base.

The choice of the key used for blocking is the most important decision in the blocking process [32]. The key must be carefully selected to provide a good reduction in the number of comparisons, while at the same time ensuring the process does not miss any possible matches. If the number of blocks is too small, for example when choosing gender as the blocking key, the number of records in each block will be too big, resulting in many extra unnecessary comparisons. If, on the other hand, blocks are too small, such as for example selecting a passport number as the key, then potential errors in the data can cause true duplicate matches to be missed.

One disadvantage with this blocking procedure is that typing errors in the key will result in potentially matching records being missed, since these records are separated into different blocks. The number of missing records in a field should also be taken into consideration. If a field has many missing values, then records with those missing values are not going to form part of any block, reducing the likelihood of matching duplicates.

To mitigate the effect of typing errors, phonetic encodings or string functions can be applied to the chosen keys [20]. A substring function that extracts the first character from the name field, can place all the names starting with that letter in the same block. Phonetic encodings convert the string of characters into a code representing the pronunciation of the word. By definition such encodings are language dependent, therefore selecting the right encoding is dependent on the language. The oldest and best known English based phonetic encoding is Soundex [42]. Soundex converts a string into the first character of the string and a set of numbers according to an encoding table. Phonex [41] and Phonix [33]are two variations on the Soundex, that attempt to improve the encoding scheme by applying more transformation to the words.

In order to evaluate blocking algorithms the measures of *pair completeness* and *reduction ratio* [8] are typically used. Pair completeness measures the number of the identified pairs by the algorithm compared with the true number of duplicates that exist in the whole dataset.The reduction ratio measures the reduction in the number comparisons when using the blocking algorithms.

4.1.3 Field Comparison

After the blocks of records have been identified, the record pairs need to be compared to determine the similarity between pairs. Depending on the classification algorithm used to classify the records, the output of each field comparison can be binary or a continuous measure of distance, typically between 0 and 1.

Functions for comparison depend on the type of data contained in the fields. Classically, most of the data in an entity resolution process involves string data, so often string distance algorithms are used to measure the similarity of fields [44].

Both Christen and Cohen et al [25] and [20] studied string matching functions in the context of name matching for entity recognition. All entity recognition processes involving person entities typically contain personal name fields that have to be compared. Personal names can have different characteristics from general text, such as multiple spellings for the same name, initial and middle name abbreviations and shortened names. The variation in name spelling can be considered as a special case of misspelling, however sometimes names change completely with name shortening. Generic string comparison algorithms typically don't cater for the worst of these cases. More complex multi-lingual cases such as John being used interchangeably with Sean(Irish) or Jean(French) are often simply ignored.

One of the best name matching algorithms, in terms of performance and robustness, identified by both Cohen and Christen in their separate studies is the Jaro-Winkler algorithm [48]. This is an extension of the algorithm Jaro proposed in [38]. The Jaro-Winkler algorithm starts with the computation of the Jaro measure then adjusts the value to give more similarity in the prefix and reduce the disagreement value of characters that are similar, such as "1" and "l".

Field comparison functions for numeric values are not as advanced as those for string functions [30]. Numeric fields can be treated as strings and compared using string distance functions. Alternatively the percentage difference between the fields can be used to quantify a normalised difference [22] measure.

4.1.4 Classification

Once field comparison is complete, each pair of records has to be classified as either a match or a non-match. The first approaches to classification came from the statistics community and relied on probability theory, to estimate the probability of a record being a match or otherwise. Felligi and Sunter [32] contributed to two main aspects of entity recognition; the calculation of field weights based on the information quality of the field, and the definition of thresholds to classify record pairs into three classes.

Before records can be classified, potentially identical records need to be compared based on their fields, however not all fields contribute equally to the final decision of whether a pair is a match or not. For example, a match on identical names is usually quite significant in identifying matching records, however a match on the date of birth or sex of a person can be less significant. In order to quantify the importance of the field, each field can be weighted according its importance, with more representative fields having a higher weight. To determine the field importance, Felligi and Sunter proposed the use of two probabilities m and u, that determine the agreement and disagreement weights of the individual fields.

Once the comparison of each of the attributes is complete, the total agreement weight can be calculated to determine the value of the weight vector. To classify the records into the three different sets, two cutoff thresholds must be defined. The *upper* threshold defines all the pairs that are matches, the *lower* threshold defines pairs that are not matches, and the records that fall in between are possible matches that could be manually reviewed if necessary. In practice, the two thresholds can be determined empirically based on the specific data set.

The probabilistic model of Fellegi and Sunter was subsequently revised and improved by other researchers [38, 57]. Subsequent approaches used rule bases written with the help of domain experts to classify records [34]. Elmagarmid et al [30] provide a comprehensive overview of the individual classification algorithms that fall into the above broad classes.

Availability of training data and advances in machine learning brought about the use of machine learning techniques to tackle the problem. The current state of the art in classification [21] uses Support Vector Machines (SVMs) for training models and classifying records, when training examples are available. SVMs have been successfully applied to several classification domains such as handwriting recognition, classifying facial expressions and text categorisation [18]. Originally, SVMs were designed to classify binary class problems, which makes them a prime candidate for entity resolution tasks, where the goal is to divide record pairs into two sets of matches and non-matches.

4.2 Relational Entity Resolution

Relational entity resolution approaches require that the data already has an inherent relational structure. These approaches exploit this relational structure to add more information to the entity resolution process to improve the classification accuracy. In the research surveyed here, relational information always improves the entity relationship accuracy when compared to attribute only techniques.

The simpler relational entity resolution techniques treat relational information as just another attribute between pairs. These approaches are based on the attribute resolution process but some of the attributes contain relational information. Relationship information is added to the comparison vector and if two records share the same relationship then the similarity of that attribute is a true match. Ananthakrishna et al [3] describe a database centric approach that exploits data hierarchies in the database as additional relational information. This information is also used to reduce the number of comparisons during the entity resolution process.

Bhattacharya and Getoor [12] describe a more complete relational model with their collective entity resolution approach. They define the entity resolution problem as a clustering problem where each cluster represents a unique entity. Clusters are merged based on their similarity which is calculated with a similarity measure that combines relational similarities and attribute similarities. The authors have shown that this approach improves both on attribute based entity resolution and on techniques that treat relationships as attributes.

4.3 Evaluation

Traditionally, information retrieval evaluation of accuracy, precision, recall and a combined f-measure score, have been used to evaluate the quality of an entity resolution process [4]. Christen and Goiser provide a comprehensive overview of the main quality measures used in entity recognition [23].

In entity resolution, it is common that there is a disproportionate ratio between the number of matches and the number of non-matches in a data set. True negatives typically occupy the vast majority of the results and if one were to blindly classify all matched pairs as negatives high scores of accuracy can still be achieved. For this reason in the case of unbalanced data sets any measures that involve a measure of true negatives should be avoided [54].

Pairwise attribute matching techniques often apply transitive closure as a final step in entity resolution. It is important to evaluate any entity resolution results after transitive closure is applied because this step can propagate error within the data. If object a is the same as b and b the same as c, the transitivity relationship will conclude that a is the same as c. If b and c are not really the same passenger than this error will duplicate itself when a is joined with c.

While the f-measure metric attempts to combine the values of precision and recall into one measure, Bilenko and Mooney [14] warn against using this measure in favour of precision recall curves. In single value measures, the measures do not provide any indication of where the cutoff threshold that separates matches from non-matches is. On the other hand, precision values interpolated at standard recall levels can highlight the performance of a classifier at different cutoff thresholds.

5 Identifying Airline Customers Case Study

In our case study we describe the extraction process of a social network of passengers travelling with an airline that was inferred from a source of passenger booking data. The data set used for this study consists of a total of 9,468,460 one-way flight passenger records from which 2,968,282 unique passengers are extracted.

The primary source of data in the sale of an airline ticket is the passenger name record (PNR). Each airline computer reservation system (CRS) has its own PNR record format, however all PNRs have a similar structure and contain approximately the same information. The PNR record contains all the information required to make a booking and buy a ticket, including the travelling passenger names, flight itinerary, passenger contact details and information on the entity that made the sale. The passenger contact details can include mail address, email addresses and phone numbers. However, only the phone number is strictly compulsory. The amount of available data is usually dependent on the source of the booking. In some cases, such as website bookings, the front-end application can make certain fields compulsory, even though the back-end CRS does not. The booking and ticket information provides a wealth of data that can be mined to provide better business intelligence and support decision making.

5.1 Identifying Customers

Whenever airlines need to analyse customer data usually the only source of data available is the frequent flyer system, which only contains passengers who voluntarily register for the frequent flyer program. A member of the frequent flyer program can be a valuable customer or can be a regular passenger. What frequent flyer membership provides is the facility to track and measure the value of a customer. Typically valuable customers eventually become members of the frequent flyer program, because of the added benefit the program gives them, however not all do.

In this context therefore it is important to distinguish between passengers and customers. A customer is a passenger who provides value to the airline. Presently customers can only be frequent flyers because the rest of the people who are not frequent flyers are just passengers, treated simply as numbers. What the concept of identifying customers refers to in this section is the identification of a valuable customer from the whole available set of passengers irrespective of whether they are frequent flyers or not.

Identifying customers first entails that each passenger that has ever made a booking with the airline is identified with a unique number. The definition and measurement of the value of the customer can be defined by the airline business analyst, however now the analyst is not restricted to members of the frequent flyer program alone. This value of a customer can be measured in different ways, amongst them travel frequency, revenue generated and the number of social ties the passenger has. Identifying potential valuable customers from the whole spectrum of passengers can prove beneficial for the airline to identify previously unknown potential valuable customers and build lasting profitable relationships with them.

Apart from targeting valuable customers to join the frequent flyer program, this information can be easily used to improve current customer support. A common example is when a passenger forgets to provide his frequent flyer number at the time of booking a flight. The system can recognise that the passenger is already a frequent flyer and interact with the frequent flyer system to notify it of the sale without the intervention of the customer, thus providing a better level of service.

5.2 *Features of the Data*

In order to study the extent of missing data which is not compulsory, a sample of over 200,000 records was analysed. Since we are concerned with uniquely identifying entities, therefore passengers, the fields of interest were mainly those that contain the passenger contact details. This data sample is specific to one airline, however airlines that operate on a similar business model have a similar distribution. Table 4 shows the results of this analysis.

Table 4 Missing records in each field

Field	Percentage Missing
Address	39.9%
Zip Code	39.9%
Frequent Flyer No	65.4%
Phones	75.8%
Email Address	78.6%
Title	93.7%
Group Name	95.3%

The only data element that uniquely identifies a passenger is the frequent flyer number, however only 35% of the passengers in dataset have a frequent flyer number. Passengers usually find value in enrolling in frequent flyer programs because they travel frequently. In order to determine the network of all the airline passengers, the remaining 65% of the passengers have to be uniquely identified.

5.3 Entity Resolution Process

For our case study we use an attribute based approach to entity recognition. The overall aim of this research is to eventually infer large networks from data that has no explicit network links or where links can be ambiguous. In this scenario, node identification is the first stage towards identifying the network structure. Current relational network approaches require the network to be already available to disambiguate between nodes. Once the network is identified the network information can be fed back to a second entity resolution pass to improve the accuracy of entity recognition and then again the accuracy of the generated network. In future work we plan to embed this stage in the network inference process and study in depth the interplay between the relationship inference and relational entity resolution.

Each of the the four stages described in section 4.1, data standardisation, blocking, field comparison, classification, involve design decisions that influence the efficiency and the outcome of the entity resolution process. In this section we will look at the design decisions we considered during this process, and the results of the most efficient and effective solution used for identifying passengers.

To facilitate the development of this procedure the Febrl framework [22] and toolkit were used. Febrl packages all the stages of the ER process in an easily extendable and customisable open source toolkit, written in Python. Originally, Febrl was developed as a research platform to assist with medical record linkage, however the generic framework made it straightforward to adapt it and use it to identify duplicate airline passengers here.

5.4 Data Standardisation and Cleansing

All the data elements extracted from the booking were sanitised to ensure processing consistency. The four main data elements that can be used to uniquely identify a passenger are the contact details available in the booking. The contact details include the frequent flyer number, email addresses, phone numbers and a single mail address. The email addresses and the mail address are not linked with the individual passengers but with the whole booking, therefore all records in the same booking had the same mail and e-mail addresses.

Apart from the personal contact details of the passengers, additional information on the passenger's route travelled was added. Frequent travellers tend to travel on the same routes multiple times, therefore this information can be used to improve the identification of the same passenger. The route information can be represented in different ways, for instance it can be represented by flight type, flight distance, or the starting point of origin of the journey.

As with any other real data source, the information contained in the booking can be incorrect or misleading. For example, a booking can be made by a second person on behalf of the person travelling, giving his contact details instead of the travelling passenger's contact details. The main purpose of using this information is to uniquely identify the passengers, rather than direct marketing and contact, therefore as long as the information is consistent, this information can still be used. Should the approach be extended to direct marketing, then stricter rules should be applied to further clean the data.

5.5 Blocking

The data set used for entity resolution consists of over 9 million name records. If one were to blindly compare all the 9 million records against each other in a cross product, the process will involve over 7.9^{12} comparisons, most of which will be unnecessary. The only two fields that could effectively be used for blocking are the name and surname, since all other fields contained many missing elements that made them unsuitable as blocking keys (see Table 4).

Field blocking with encodings and sorted neighbourhood blocking were tested in the blocking stage. Three phonetic encodings were tested; Soundex, Phonex and Phonix. For these encodings, the name and surname were independently encoded phonetically, then concatenated together. The best phonetic encoding according to our tests was the Soundex encoding as it has both the highest pair completeness, and the highest reduction ratio (see Figure 7).

The sorted neighbourhood approach was explored with two different window sizes. The accuracy was only slightly better than the Soundex encoding of name and surname combined, however the number of pairs compared was significantly greater. The sorted neighbourhood approach is more efficient when several keys are used to define multiple blocks, which are then combined together [34]. Since in this scenario the number of possible keys for blocking is limited, the sorted neighbourhood approach could not be applied to its full potential.

Further experiments were held to improve the efficiency of the blocking procedure for this particular data set. The best result was achieved by using a Soundex encoding of the surname concatenated with the first two characters of the name. This approached reached a pair completeness of 99%, resulting in less than 100 actual records missed.

Figure 7 compares the efficiency of the different blocking types. The first three bars are for the concatenated name and surname with Soundex, Phonex and Phonix encodings. The next two are for the Soundex encoded surname and the first or second character of the name. The last measurement is for the sorted neighbourhood approach with a window size of 10. For all the measures the reduction ratio was over .999 of all the number of possible comparisons.

The records that are missed with this blocking approach are mainly due to passengers changing their surname after marriage. Using the name and surname keys alone makes this case very difficult to identify automatically. Using any part of the surname is always prone to this problem, however names are also prone to abbreviations, therefore the most accurate blocking key in this case is using the first letter of

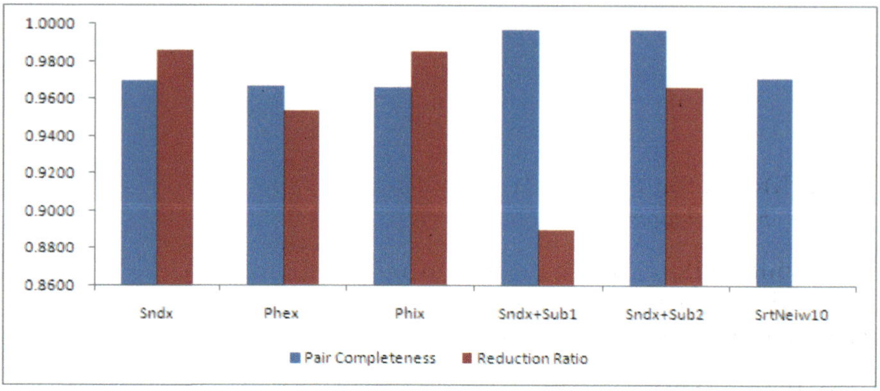

Fig. 7 Blocking Result

the name. This approach however will make the number of comparisons too high, with only 24 possible buckets, one for each letter of the alphabet. The benefit in terms of accuracy is marginal while the number of extra comparisons is significantly large.

5.6 *Weight Generation*

After the potential matching names are grouped together in the blocking stage, the weight of each pair of records was calculated to form a weight vector. The weights were calculated based on the field datatype and the content of the field. Some measurements that were used include the following:-

Jaro-Winkler: This fuzzy string comparison function defined by Jaro-Winkler [56] returns a figure between 0-1 depending on the similarity between strings.

Exact String Match: Extract string comparison. If the strings are not exactly the same then 0 is returned.

Max String Difference: This defines the maximum number of characters that can differ between strings. If the maximum number of differences is smaller than the threshold then 1 is returned.

Minimum Set Membership: If at least X members of the set are the same, where X is the set threshold then a match of 1 is returned.

Flight Boolean Match: If the number of flights > 0 in both fields, or the number of flights $= 0$ in both fields, then the function returns a match value of 0.

Numeric Percentage Difference: If the difference between the two numbers is less than the percentage threshold then 1 is returned, 0 otherwise.

The weight vectors generated during this stage have a direct effect on the quality of the generated model. Different combinations of weight calculations for the available fields were tested. The use of flight preference fields and shopping preference fields was also tested. The best resulting model included the flight preference fields, but not the shopping preference fields.

5.7 Classification

The classification stage determines if each pair of records extracted from the blocking stage and subsequently weighed in the weight generation stage, is a match or not. Three different approaches are evaluated to classify the record pairs as matches and non-matches. The first approach is the traditional approach described by Fellegi and Sunter, the second approach uses a set of declarative if-then rules and the third approach uses a supervised SVM classifier using the libsvm [19] library.

Importantly, the frequent flyer number enabled the testing and evaluation of the entire classification process. Four different record sets of random records containing frequent flyer numbers were extracted from the data set. There was no overlap between records in each data set, so each record was present in only one data set. One of the record sets was used only to train the SVM classifier, and in the case of the rule base and the Fellegi Sunter this set was used to empirically adjust the thresholds of rules and classification.

The cut-off thresholds of the Fellegi and Sunter were determined on the training set by separating the number of matches and non matches in different weight buckets and using the resulting thresholds for matches and non matches. The rules for the rule base approach were encoded according to the understanding of the data set. The application of the rules was tried several times on the testing set to determine the best group of rules and the best value threshold for the rules. After the thresholds were set the same rules were applied to the three different testing sets.

For the SVM classification the training set was used to train two types of classifiers, a linear classifier and a RBF classifier. For the training of the RBF classifier, 10 fold cross validation was used to determine the best parameters for C and γ. For the linear classifier, three different values for C (0.1, 1, 10) were used and the best parameter of the three (10) was chosen. The models generated by the SVM training were saved and subsequently applied to the three testing sets. In the training of the SVM the frequent flyer number was only used to identify a pair of records as a match or a non-match, but did not form part of the weight vector of attributes. This approach allows us to report on our results which a high degree on confidence due to the frequent flyer number ground truth data.

For each of the three classification approaches the accuracy, precision, recall and f-measure were calculated for the training set used and the three different testing sets. As discussed in section 4.3 the accuracy values for an unbalanced data set task is usually skewed because of the disproportion between matches and non matches. For this reason we based the evaluation on individual precision and recall values.

The two supervised learning models were superior to the two unsupervised models by around 20% on average in terms of precision. Figure 8 shows these results. This result was expected and confirms what other researchers have found, in that supervised learning techniques provide better results than unsupervised approaches.

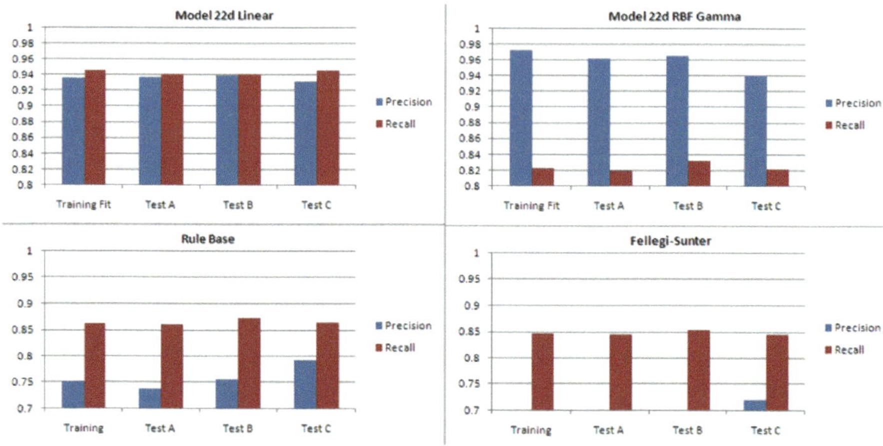

Fig. 8 Classification Results

The SVM RBF model gave the best result in the evaluation. This model was then applied to the remaining set of data that did not contain any frequent flyer information. The output of this process resulted in a list of matching record pairs. To determine all the records referring to the same passenger, the graph components were extracted from the data set and all the passengers referring to the same entity were assigned the same unique identifier.

6 Conclusion

Inferring relational information from attribute based data sets is currently one of the few ways that large scale network data can be collected. In this chapter we explored how actors in a network can be identified before extracting the relationships between them. The well studied area of entity resolution was surveyed and detailed as it provides an acceptable approach and developments in the area are continually progressing.

Actor identification is however, only the first aspect of network inference. Once the actors are identified the relationships between the actors have to be extracted. These relationships could in turn improve the accuracy of actor identification within a cyclic feedback loop. In future work we aim to integrate actor identification and relationships extraction in a common framework to infer social networks.

References

1. Albert, R., Barabási, A.: Statistical mechanics of complex networks. Reviews of Modern Physics 74(1), 47–97 (2002)
2. Albert, R., Jeong, H., Barabási, A.: Diameter of the world wide web. Nature 401(6749), 130–131 (1999)
3. Ananthakrishna, R., Chaudhuri, S., Ganti, V.: Eliminating fuzzy duplicates in data warehouses. In: Proceedings of the 28th international conference on Very Large Data Bases, pp. 586–597. VLDB Endowment (2002)
4. Baeza-Yates, R., Ribeiro-Neto, B.: Modern information retrieval. ACM Press, New York (1999)
5. Barabási, A., Oltvai, Z.: Network biology: understanding the cell's functional organization. Nature Reviews Genetics 5(2), 101–113 (2004)
6. Bausch, S., Han, L.: Social Networking Sites Grow 47 Percent, Year Over Year, Reaching 45 Percent of Web Users, According to Nielsen. NetRatings, Nielsen/Netratings, press release 11 (2006)
7. Baxter, I., Quigley, A., Bier, L., Moura, L., Sant'Anna, M.: Clonedr: Clone detection and removal. In: 1st International Workshop on Soft Computing Applied to Software Engineering, SCASE 1999 (1999)
8. Baxter, R., Christen, P., Churches, T.: A comparison of fast blocking methods for record linkage. In: Proceedings of the KDD 2003 workshop on data cleaning, record linkage, and object consolidation, Washington DC, vol. 3, pp. 25–27 (2003)
9. Bell, M., Iida, Y.: Transportation network analysis. Wiley, Chichester (1997)
10. Benjelloun, O., Garcia-Molina, H., Kawai, H., Larson, T., Menestrina, D., Su, Q., Thavisomboon, S., Widom, J.: Generic entity resolution in the serf project. IEEE Data Engineering Bulletin 29(2), 13–20 (2006)
11. Berry, M.J., Linoff, G.: Data mining techniques: for marketing, sales, and customer support. John Wiley & Sons, Inc., New York (1997)
12. Bhattacharya, I., Getoor, L.: Collective entity resolution in relational data. ACM Trans. Knowl. Discov. Data 1(1), 5 (2007), `http://doi.acm.org/10.1145/1217299.1217304`
13. Bilenko, M., Basu, S., Sahami, M.: Adaptive product normalization: Using online learning for record linkage in comparison shopping. In: ICDM 2005: Proceedings of the Fifth IEEE International Conference on Data Mining, pp. 58–65. IEEE Computer Society, Washington (2005), `http://dx.doi.org/10.1109/ICDM.2005.18`
14. Bilenko, M., Mooney, R.J.: On evaluation and training-set construction for duplicate detection. In: Proceedings of the KDD 2003 workshop on data cleaning, record linkage, and object consolidation, Washington DC, pp. 7–12 (2003)
15. Bilgic, M., Licamele, L., Getoor, L., Shneiderman, B.: D-dupe: An interactive tool for entity resolution in social networks, pp. 43–50 (2006), doi:10.1109/VAST.2006.261429
16. Boyd, D.M., Ellison, N.B.: Social network sites: Definition, history, and scholarship. Journal of Computer-Mediated Communication 13(1) (2007)
17. Van de Bunt, G., Van Duijn, M., Snijders, T.: Friendship networks through time: An actororiented dynamic statistical network model. Computational & Mathematical Organization Theory 5(2), 167–192 (1999)
18. Burges, C.: A tutorial on support vector machines for pattern recognition. Data mining and knowledge discovery 2(2), 121–167 (1998)
19. Chang, C.C., Lin, C.J.: LIBSVM: a library for support vector machines (2001), Software available at, `http://www.csie.ntu.edu.tw/~cjlin/libsvm`

20. Christen, P.: A comparison of personal name matching: Techniques and practical issues. Tech. Rep. TR-CS-06-02 (2006)
21. Christen, P.: Automatic record linkage using seeded nearest neighbour and support vector machine classification. In: KDD 2008: Proceeding of the 14th ACM SIGKDD international conference on Knowledge discovery and data mining, pp. 151–159. ACM, New York (2008), http://doi.acm.org/10.1145/1401890.1401913
22. Christen, P., Churches, T., Hegland, M.: Febrl-a parallel open source data linkage system. Lecture notes in computer science pp. 638–647 (2004)
23. Christen, P., Goiser, K.: Quality and complexity measures for data linkage and deduplication. Quality Measures in Data Mining 43, 127–152 (2006)
24. Churches, T., Christen, P., Lim, K., Zhu, J.: Preparation of name and address data for record linkage using hidden Markov models. BMC Medical Informatics and Decision Making 2(1), 9 (2002)
25. Cohen, W.W., Ravikumar, P., Fienberg, S.E.: A comparison of string distance metrics for name-matching tasks, pp. 73–78 (2003)
26. Collberg, C., Kobourov, S., Nagra, J., Pitts, J., Wampler, K.: A system for graph-based visualization of the evolution of software. In: Proceedings of the 2003 ACM symposium on Software visualization. ACM, New York (2003)
27. Domingos, P., Richardson, M.: Mining the network value of customers. In: Proceedings of the seventh ACM SIGKDD international conference on Knowledge discovery and data mining, pp. 57–66. ACM, New York (2001)
28. Dunn, H.: Record linkage. American Journal of Public Health 36(12), 1412 (1946)
29. Eagle, N., Pentland, A., Lazer, D.: Inferring Social Network Structure using Mobile Phone Data. PNAS (2007)
30. Elmagarmid, A., Ipeirotis, P., Verykios, V.: Duplicate Record Detection: A Survey. IEEE Transactions on knowledge and data engineering, 1–16 (2007)
31. Farrugia, M., Quigley, A.: Enhancing airline customer relationship management data by inferring ties between passengers. In: Proceedings of the international conference on Social Computing (2009)
32. Fellegi, I., Sunter, A.: A theory for record linkage. Journal of the American Statistical Association, 1183–1210 (1969)
33. Gadd, T.: PHONIX: The algorithm. Program–Electronic Library and Information Systems 24(4), 363–366 (1990)
34. Hernández, M., Stolfo, S.: Real-world data is dirty: Data cleansing and the merge/purge problem. Data Mining and Knowledge Discovery 2(1), 9–37 (1998)
35. Hill, S., Provost, F., Volinsky, C.: Network-based marketing: Identifying likely adopters via consumer networks. Statistical Science 21(2), 256 (2006)
36. Hirschman, L., Chinchor, N.: Muc-7 coreference task definition - version 3.0 (1997)
37. InfoGlide Software: Fighting workers' compensation fraud using identity recognition. Tech. rep., InfoGlide Software (2009)
38. Jaro, M.: Advances in record-linkage methodology as applied to matching the 1985 census of Tampa, Florida. Journal of the American Statistical Association, 414–420 (1989)
39. Krackhardt, D., Hanson, J.: Informal networks: the company. Knowledge in Organizations (1996)
40. Krebs, V.: Mapping networks of terrorist cells. Connections 24(3), 43–52 (2002)
41. Lait, A., Randell, B.: An assessment of name matching algorithms. Technical Report Series-University of Newcastle Upon Tyne Computing Science (1996)
42. Odell, M., Russel, R.: The soundex coding system. US Patent (1918)
43. Marsden, P., Campbell, K.: Measuring tie strength. Social Forces 63, 482 (1984)

44. Navarro, G.: A guided tour to approximate string matching. ACM Computing Surveys (CSUR) 33(1), 31–88 (2001)
45. Newcombe, H., Kennedy, J., Axford, S., James, A.P.: Automatic linkage of vital and health records. Science 130, 954–959 (1959)
46. Petróczi, A., Nepusz, T., Bazsó, F.: Measuring tie-strength in virtual social networks. Connections 27(2), 39–52 (2006)
47. Piatetsky-Shapiro, G., Djeraba, C., Getoor, L., Grossman, R., Feldman, R., Zaki, M.: What are the grand challenges for data mining. KDD-2006 Panel Report. SIGKDD Explorations 8(2), 70–77 (2006)
48. Porter, E., Winkler, W.: Approximate String Comparison and Its Effects on an Advanced Record Linkage System. U.S. Bureau of the Census, Statistical Research Division (1997)
49. Rahm, E., Do, H.: Data cleaning: Problems and current approaches. IEEE Data Engineering Bulletin 23(4), 3–13 (2000)
50. Scharstein, D., Szeliski, R.: A taxonomy and evaluation of dense two-frame stereo correspondence algorithms. International Journal of Computer Vision 47(1), 7–42 (2002) Has 1205 Citations
51. Scott, J.: Social Network Analysis: A Handbook, 2nd edn. SAGE Publications, Thousand Oaks (2000)
52. Sole, R., Murtra, B., Valverde, S., Steels, L.: Language Networks: their structure, function and evolution. Trends in Cognitive Sciences (2006)
53. Statistics New Zeland: Data Integration Manual (2006), `http://www.stats.govt.nz/NR/rdonlyres/35662748-4DBC-41DA-A519-E6D9D7748C20/0/DataIntegrationManual.pdf`
54. Tan, P., Steinbach, M., Kumar, V.: Introduction to Data Mining. Addison-Wesley, Reading (2005)
55. Wasserman, S., Faust, K.: Social network analysis: Methods and applications. Cambridge Univ. Pr., Cambridge (1994)
56. Winkler, W.: String Comparator Metrics and Enhanced Decision Rules in the Fellegi-Sunter Model of Record Linkage. pp. 354–359 (1990)
57. Winkler, W.: Improved decision rules in the fellegi-sunter model of record linkage. In: Proceedings of the Section on Survey Research Methods, pp. 274–279. American Statistical Association (1993)
58. Winkler, W.: The state of record linkage and current research problems. Statistical Research Division, US Bureau of the Census, Wachington, DC (1999)
59. Zachary, W.: An information flow model for conflict and fission in small groups. Journal of Anthropological Research, 452–473 (1977)

Perception of Online Social Networks

Travis Green and Aaron Quigley

Abstract. This paper examines data derived from an application on Facebook.com that investigates the relations among members of their online social network. It confirms that online social networks are more often used to maintain weak connections but that a subset of users focus on strong connections, determines that connection intensity to both connected people predicts perceptual accuracy, and shows that intra-group connections are perceived more accurately. Surprisingly, a user's sex does not influence accuracy, and one's number of friends only mildly correlates with accuracy indicating a flexible underlying cognitive structure. Users' reports of significantly increased numbers of weak connections indicate increased diversity of information flow to users. In addition the approach and dataset represent a candidate "ground truth" for other proximity metrics. Finally, implications in epidemiology, information transmission, network analysis, human behavior, economics, and neuroscience are summarized. Over a period of two weeks, 14,051 responses were gathered from 166 participants, approximately 80 per participant, which overlapped on 588 edges representing 1341 responses, approximately 10% of the total. Participants were primarily university-age students from English-speaking countries, and included 84 males and 82 females. Responses represent a random sampling of each participant's online connections, representing 953,969 possible connections, with the average participant having 483 friends. Offline research has indicated that people maintain approximately 8-10 strong connections from an average of 150-250 friends. These data indicate that people maintain online approximately 40 strong ties and 185 weak ties over an average of 483 friends. Average inter-group accuracy was below the guessing rate at 0.32, while accuracy on intra-group connections converged to the guessing rate, 0.5, as group size increased.

Keywords: social network analysis, social network perception, node proximity.

Travis Green
Cognitive Science Masters Programm, University College Dublin, Ireland
e-mail: `travis.h.green@gmail.com`

and

Research Associate, Neukom Institute for Computational Sciences, Dartmouth College, Hanover, NH, USA

Aaron Quigley
CASL, University College Dublin, Belfield, Dublin 4, Ireland
e-mail: `aquigley@ieee.org`

I.-H. Ting et al. (Eds.): Mining and Analyzing Social Networks, SCI 288, pp. 91–106.
springerlink.com

1 Introduction

Social networks enable humans to gain access to information, support in time of need, and form an immensely important part of our everyday lives. From business cards and address books to firm handshakes and attentive listening, a significant part of our actions are devoted to creating and maintaining our networks.

Over the past two decades, those connections have become increasingly supplemented by online media, beginning with message boards, progressing to email and instant messaging, and now blooming into online social networking sites such as Facebook.com ("Facebook").

Research into our online social networks is in its infancy. This paper will discuss recent relevant research, describe this study and its results, and focus on implications for broader fields.

2 Background

2.1 Definition and History

Boyd and Ellison define online social network sites as websites that enable users to construct a profile describing themselves, show who they feel connected to, and view the connections of others [1]. Many users join with the goal of demonstrating their social networks, connecting to a larger extended network, gaining access to job or travel opportunities, and communicating with geographically distant friends [1], [2].

Online social networking is a recent phenomenon, with most scholars tracing its inception to SixDegrees.com in 1997 [1]. A second generation of networks developed focused on specific communities, such as job-seekers and ethnic community. Developers believed that by addressing the needs of a single community, they hoped to focus on common interests, as Feld [3] suggests, and cross link those communities to create higher levels of interaction. Many of these networks succeeded in gaining their niche communities, but soon encountered new obstacles that prevented mass adoption, as users left the sites when their bosses, friends, and family were all presented with the same online persona.

Facebook is the leading social networking service on the Internet today, with billions of page views per day making it the sixth most visited website, and a user base now numbering close to 200 million [4],[5], [6]. Users can upload a profile and pictures, make links to "friends," post public messages for others, and add applications that enhance their online experience [2]. Users are able to view the profiles of their friends after both confirm their a desire to be connected, and the connections that their friends have made to others [1]. The typical user accesses Facebook heavily, logging in for 20 minutes a day on average with two thirds of all users logging in at least once each day [5].

Social networks enable us to connect digital data to our offline society, but can require a significant investment of time and energy while possibly giving others an inaccurate view of ourselves.

2.2 *Connection Strength*

Sociologists studying social networks in practice have focused on the use of strong ties compared to weak ties. Wellman observed that, despite cultural variability in quantity, people categorize friends into three approximate groupings: acquaintances, active contacts, and intimate friends, with a vast drop off in numbers as intimacy increases [7].

Strong ties, also known as bonding ties, provide companionship and support, and tend to form small and tightly linked networks of friends [8]. Close friends and families provide excellent social support, but require significant investment to maintain, and tend to provide new information infrequently. Recent studies from Facebook show that commenting and messaging indicate maintenance of 10-26 strong online ties [9]. However, messaging may not indicate closeness, just as a physical connection does not always result in an online connection.

Weak ties, or bridging ties, enable significant diversity of information and opportunities to be collected [8]. Granovetter [10] defines the strength of a social tie to be a "combination of the amount of time, the emotional intensity, the intimacy (mutual confiding), and the reciprocal services which characterize a tie." Usually based on a specific context, these networks are low maintenance, and enable the collection of novel information and opportunities because they represent connections to more diverse clusters [8]. Similarly, information transmission is enhanced by favoring weak ties, as such ties will ensure that the information does not become trapped within cliques [10].

Unlike strong ties, the numbers of weak ties appear to be enhanced significantly by digital technology by reducing the cost of maintaining these connections [8]. Since more weak tie connections can be made, social networks become more valuable as available information and opportunities are increased [8]. Ideally one wants to bridge disparate groups as then information from both is available, and the person bridging gains status from the connections they could make.

Ties of all strengths appear to be enhanced by online social networks. Online networks show growth patterns similar to measurements made offline, indicating that inferences from physical world studies can be carried into the digital realm. As shown in this study, some online connections are more important to individuals than others, creating a direct parallel to these offline categories.

2.3 *Social Network Perception*

Freeman *et al.* [11] examined a tightly knit windsurfing community and show that members are highly accurate at reporting observed general patterns of association, but remember specific instances poorly. Janicik and Larrick [12] describe two well-established mental schemas observed in human attempts at understanding surrounding social relationships. The balance schema assumes that friendships are reciprocal and transitive. The linear-ordered schema assumes that influences can be asymmetric and transitive. This means that any friendship relationship is two-way, and that information can spread outward through the network from its source [12]. These schemas make us more likely to perceive that a missing relation exists than to assume that an existing relation is missing. They further postulate that high-performing individuals have learned template patterns that they rapidly apply to learn the presented patterns, indicating that performance may be trainable [12]. Both online and physical network studies indicate similar perceptions of external networks, with consistent biases towards reciprocity, transitivity, and outward spread of information.

From this research, we would expect that users will have a predisposition to assume that their close friends are connected offline and online, and that users assume that the presence of an offline connection indicates the presence of an online one, a false positive.

3 Preliminary Study

The preliminary study was intended to refine techniques to investigate our four primary hypotheses:

(1) online social networks contain larger numbers of weak connections that represent an increased diversity of information flow,
(2) connections rated to have higher intensity will be perceived more accurately,
(3) increased user closeness correlates with improved perceptual accuracy about social networks, and
(4) connections within a completely connected subcomponent, or clique, will be more accurately perceived than those between cliques.

To prove these hypotheses, data needed to be gathered regarding an overall pattern of participant perception as a control, data which focused on overlapping connections with other participants for comparison, and targeted questions focused on examining connections within their social subgroups and between those groups. The preliminary study enable refinement of techniques to approach these problems.

3.1 Application Design

A Facebook application, "FriendOracle," (apps.facebook.com/friendoracle) was constructed to gather the necessary data on the underlying social network, and to select the most information-rich questions to ask individual users. This application used the scripting language PHP to integrate with Facebook's proprietary databases and a local MySQL database. These processes were abstracted out of the user experience using Facebook's HTML scripting language, FBML, to present the user with three pages:

Home – described the purpose of the study, and presented minimal user statistics

Train – presented twelve questions of the format: "Does {User A} know {User B}?" and gave users options to select the strength of the given connection {1,2,3,4,5} and the type {Facebook

Only, Real World Only, Both, Neither, Neither but they should be!}

Stats – presented detailed user statistics including comparisons among users, a response to the ideas of beta users in the preliminary study.

Once users had completed a given set of questions, they were presented with a table comparing their answers with Facebook's data, and score and accuracy metrics. As this page was generated, their results were archived into the dataset.

A Facebook application was chosen over other forms of study because it automatically affords access to a real-time sampling of the participants' social network that contains connections ranging from acquaintances to close friends. While many likely connections can be inferred from other users' perceptions, an important limitation of this study is that Facebook networks are inherently incomplete pictures of one's social network. However, the quality and breadth of already available data greatly simplifies establishing a picture of a participants' social network, a key issue with previous real-world studies. Lewis *et al.*[4] discuss these advantages and include: its inherent real world nature, full population description, and complete demographic information. Simultaneously, participant recruitment is not limited to a given social group, but rather can spread organically outward as participants recommend the study to their friends, and can be undertaken at any computer across the world.

Data points were selected to form points of comparison between users, evenly split between self-ratings, those where the user rates their own connections, present connections, and missing connections as summarized in pseudocode below:

```
//Self-ratings
     Select all current friends using application
     Choose randomly among them if too many
     Fill with random sample of friends if insufficient
//Present/Absent Connections (selected similarly)
     Select all overlapping edges (other participants) where
        connection present or absent
     Order by whether have self-rated connection to either user in
        edge
     Add some random edges
//Selection
     Seed a list evenly with three element groups
     Select half to create random distribution
```

This algorithm optimizes overlap between both the participants' own connections and with other participants.

3.2 User Survey

In order to refine the user experience and the effectiveness data selection algorithm, a preliminary usability study was conducted consisting of a pre-use questionnaire, exposure to the application, and a post-use questionnaire. The initial application contained all basic features excepting the *Stats* page. Overall comments were positive, with eight of ten users saying they enjoyed the experience and would play again.

Pre-Use questions focused on the size and composition of participants online and offline social networks, and how they use social networking software. All participants thought they knew their social networks well, and "usually" become friends on Facebook with people they meet offline, and the vast majority "almost never" meeting people offline they meet on Facebook. Estimates of communication by strength and type, including email and Facebook, showed little correlation among users.

Most users completed five to ten surveys in the allotted ten minute timeframe. From these results, a desired participant response of 200 questions was selected under the assumption that this selection would take approximately 15-20 minutes.

After using the software, eight of ten users said they would use the software again, with the remainder questioning the utility of rating one's friends, a manageable loss percentage for the study. Approximately half of respondents requested the ability to compare their results with their friends. Users found the selected data points interesting, and performed exceedingly well, with many reporting 100% accuracy, and indicating that the questions needed to be harder.

From these user comments, a series of modifications were made focusing on an improved participant experience and improving data collection methods. As many users requested, comparisons to one's friends was added in the form of the *Stats* page, containing groupings of "People Who Know You Best" and "People You Know Best." Some users mentioned that the pictures contained in the quiz were too small to be helpful, and their size was significantly increased. The selection algorithm was also modified to better incorporate the presence of associated downloads in the dataset. Some users were also found to log in once, complete a quiz, and log out assuming that those twelve questions were sufficient. Therefore, a progress bar was added to indicate to users how much data remains to be collected from both the individual user and overall.

From this study, the selection algorithm and user interface were vastly improved.

4 Results

Participant responses indicate that users have proportionately more friends of weaker connection strength, are more accurate at gauging connection intensity when the connection is strong and they claim to know both participants well, and connections within cliques are perceived more accurately than those outside. These results confirm the primary hypotheses being tested:

(1) online social networks contain larger numbers of weak connections that represent an increased diversity of information flow,
(2) connections rated to have higher intensity will be perceived more accurately,
(3) increased user closeness correlates with improved perceptual accuracy about social networks, and
(4) connections within a completely connected subcomponent, or clique, will be more accurately perceived than those between cliques.

4.1 General Data

Over a period of two weeks, the application gathered 14,051 individual responses from 166 participants, approximately 80 per participant, which overlapped on 588 edges representing 1341 responses, approximately 10% of the total. Participants were primarily university-age students from English-speaking countries, and included 84 males and 82 females. Responses were gathered from a random sampling of each participant's online connections, representing 953,969 possible connections, with the average participant having 483 friends. The structure of participant responses is shown graphically in Figure 1.

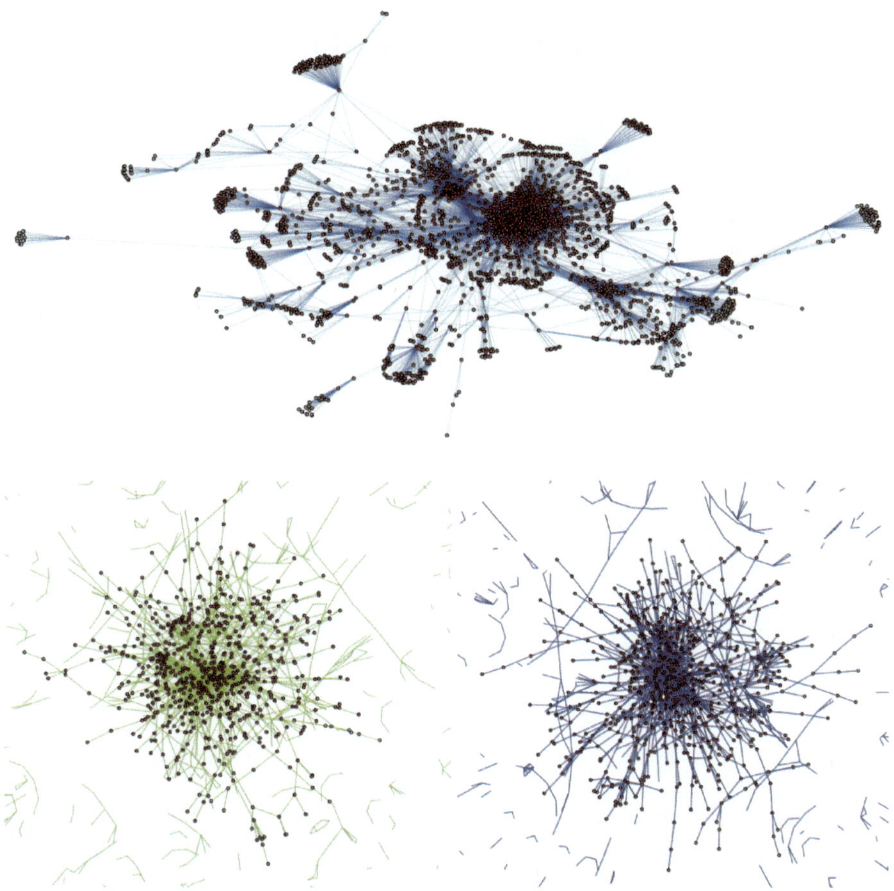

Fig. 1 (top) The network of data collected is shown. Circles represent individuals. Blue lines represent individual participant responses **(left)** Displays weak connections within the top graph with the weakest in yellow and weak in green **(right)** Displays strong connections within the top graph with the strongest as dark blue and strong as light blue

Linearly increasing friend network size correlates with polynomial growth in network connectivity. If humans have a finite cognitive capacity for memorizing connection networks, one would expect significantly reduced accuracy in perceiving friend networks. As Figure 2 shows, users with larger friend networks do display small decreases in average accuracy, but that are not significant enough to assume that the human capacity for understanding online friend networks is not adaptable. Similarly, participants' high performance even when they have many friends indicates significant intrinsic or trained ability.

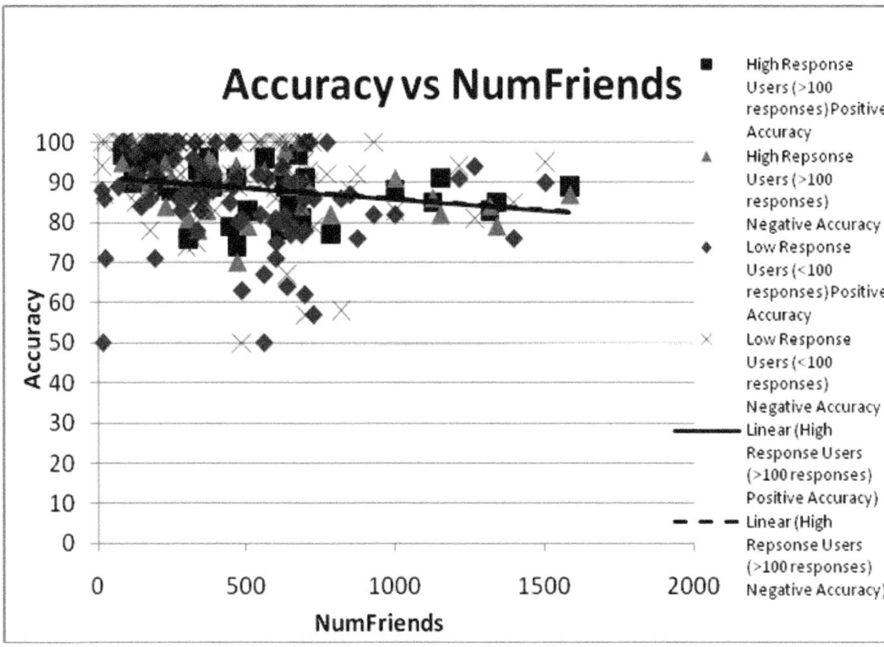

Fig. 2 Presents the relationship between a participant's number of friends on the x-axis and their overall perceptual accuracy on the y-axis. Increased number of friends is shown to correlate with decreased accuracy in both low and high response users, and on both positive and negative accuracy. (n = 166)

There exist two possible types of errors, false positives and false negatives and two possible types of correct answers, accurate positives, and accurate negatives. As Figure 3 shows, when participants claimed to be close to the two people involved in the connection in question, their rate of false negatives dropped, while their rate of false positives increased slightly. This data adds evidence to Janicik and Larrick's [12] transitivity assertion, that when the participant knows both people, a connection between them is assumed.

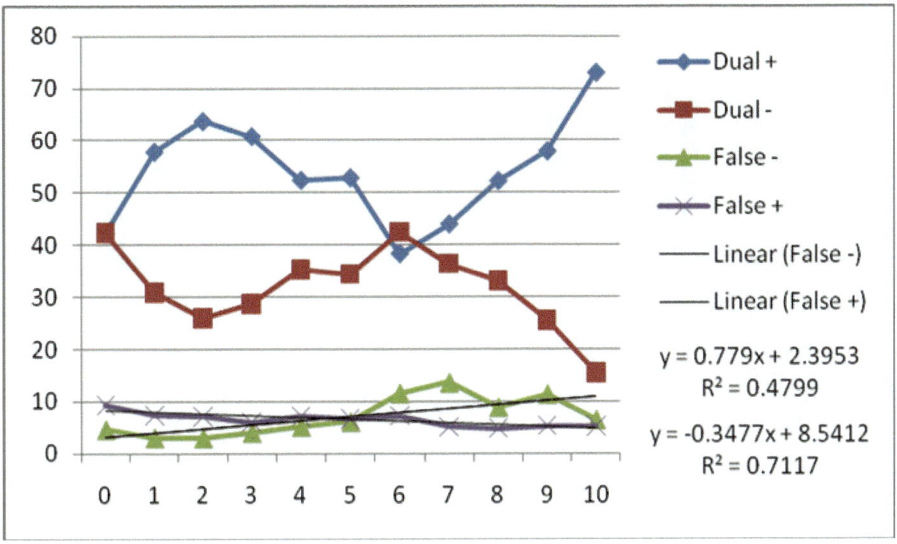

Fig. 3 Correlates the sum of observed closeness to each connected individual on the x axis and participant accuracy on the y axis. The data show evidence of an increased rate of false positives as participants increase in closeness to the connected individuals. The dip at a connection value of 10 in the rate of false positives, or a connection value of 5 to each user, may be attributed to the low number of responses fitting these criteria. (n = 14,051)

Demographic data on the user base was also collected focusing on age, birth location, sex, education, and regional affiliations. Insufficient data was collected for analysis on any metric other than sex. Hypothesis testing showed sex to be an insignificant predictor ($n = 166$, $\sigma = 13.94$, $\mu = 87.6$, $p < 0.05$).

4.2 Connection Intensity

Granovetter postulated that people maintain many more weak ties than strong ties, and Donath extended this hypothesis by claiming that online networks would enhance the number of these ties. Early candidate metrics from Facebook indicate that users may also maintain more strong ties. Offline research has indicated that people maintain approximately 8-10 strong ties from an assumed average of 150-250 friends. These data indicate that people maintain approximately 40 strong ties and 185 weak ties over an average of 483 friends. Assuming no increase in friend group size, these data indicate that online strong ties lie outside the expected range, with a proportionately smaller but absolutely larger group of weaker ties. Figure 5 describes results for each type of connection in more detail. Overall the data indicate that one's group of Facebook connections that also exist in the real world represent the strongest type of connections.

In Figure 4, N/A indicates a non-response, and should have a frequency at 0 intensity approaching 1. FB indicates a connection that is only present on Facebook,

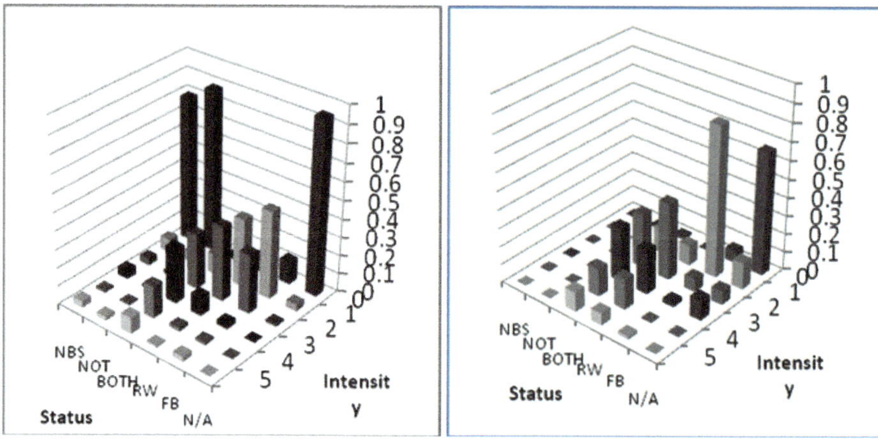

Fig. 4 (left) Shows the frequency of response scaled to 1 for each status type (x-axis) and excluding all self-ratings by participants. Intensity is described on the y-axis as ranging from 0 (nonexistent) to 5 (very strong), and frequency is represented as a percentage on the z-axis. (n = 14,051) **(right)** Shows a similar data categorization, but instead only including responses where the participant has rated one of their own connections. (n = 2,934)

which participants considered as primarily "very weak." RW indicates a connection is only present in the real world. Such connections follow a more even distribution, but one that remains heavily skewed towards weaker connections, although the mean higher than FB. BOTH represents connections present in both spheres, which are considered stronger than any other type of connection, but remain skewed towards weaker connections than a normal distribution, supporting the Granovetter [10] and Donath's [8] hypotheses. NOT represents connections that do not exist in either the real or online worlds, and should have a frequency at intensity 0 approaching 1, as is observed. NBS represents connections not present on Facebook or in the real world that the participant thinks should exist. These data conform to the Granovetter [10] and Donath [8] hypothesis that users tend to have higher numbers of connections that they consider weak.

In Figure 4(right), we see that users themselves have an increased bias towards considering Facebook-only connections to be weak, and a proportionately higher average intensity ranking for real world connections. Data in the BOTH category shows little statistical difference but maintains the bias towards higher numbers of weaker connections. This indicates that people perceive online connections to be stronger than participants consider them.

Focusing at the level of the individual participant, there does appear to be a spectrum in self-perceptions about connectivity strength. As Figure 5 shows, some users consider their Facebook connections to be much more significant than others, a result independent of friendship network size (not shown).

Figure 5 confirms Hypothesis (1), that many more weak connections exist than strong on the group level, and indicates that, we can examine user accuracy as a function of perceived strength. Hypothesis (2) states that connections perceived to

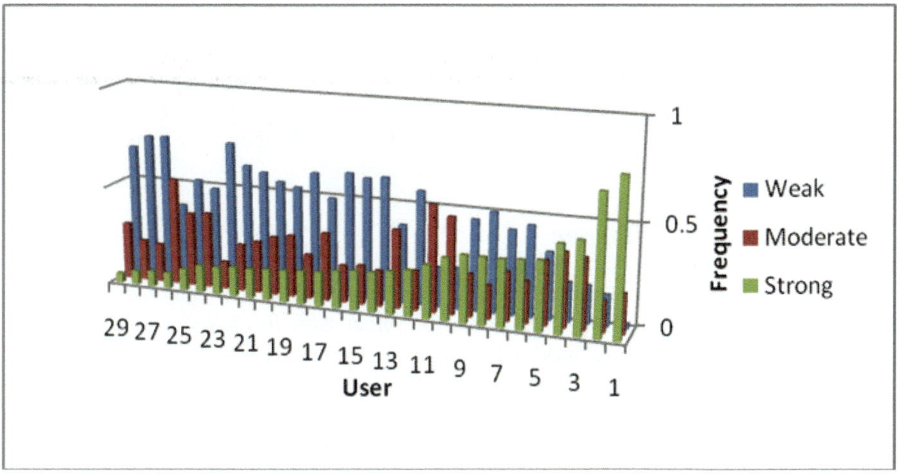

Fig. 5 Participant-level responses about their own connections categorized by enumerated intensity, limited to users providing 25 or more responses in this category. Individual users are graphed on the x-axis, and their frequency of response normalized to 1 is shown in the z-axis. A spectrum of usage is indicated, as there appear to be those focused on maintaining strong connections, and those focused on maintaining weaker ones. The presence of users indicating their strong online ties questions the intuition of Hypothesis (3), that online social networks are used primarily for maintaining weaker ties, and indicates that more research is necessary. (n = 2407)

be stronger, those rated higher, should be found more accurately by participants. Figure 6 shows this to be the case.

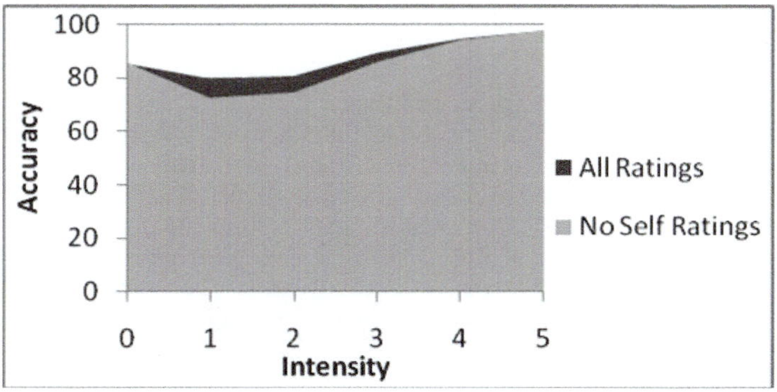

Fig. 6 Compares perceived connection intensity on the x-axis to accuracy on the y-axis. At intensity 0, here limited to a lack of connection, users are shown to be as accurate as connections rated with an intensity of 3. At intensities beyond this value, 4 and 5, user accuracy approaches 100%. (n = 14,051 ratings)

These examinations have established that this group of participants overall has more weak ties than strong ties on Facebook, although some appear focused on maintaining only strong ties, and that the overall quantity of both weak and strong ties may be enhanced by online networking technology. This section has also established that connections perceived to have a higher intensity are more accurately perceived than those with a lower intensity.

4.3 Perceptual Differences

In this section, we will examine data testing Hypothesis (3): that increased user closeness correlates with improved perceptual accuracy about social networks. We will first examine a simple model based on connections with a single user, and progress to a model that integrates connections to both users.

Figure 7 presents our simple model. It shows that connection to a single user predicts more accurately than baseline, but not that increasing connection intensity increases accuracy.

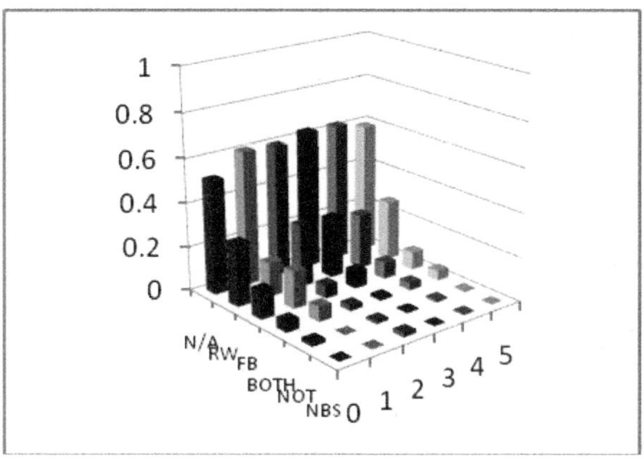

Fig. 7 Examines the accuracy of participants perceptions be comparing category (x-axis), labels described in caption of Fig 5, to absolute difference in perceived intensity (y-axis). Where available, a connection member's own rating of their connection is considered baseline, otherwise the rounded average is assumed as baseline. Under this schema, a perfectly accurate perception will be given a rating of 0, while a completely inaccurate perception will be given a score of 5. All values are scaled so that perceived intensity sums to 1. The data show that there exists a statistical advantage when a connection of arbitrary strength is present, but no advantage beyond that. (n = 518 ratings)

To integrate the user's connection to both ratings, Figure 8 follows the intuition of Figure 3 in summing the participants' perceived connection intensity to both participants in the queried connection. This approach shows a clear rise in accurate connections, defined as those scoring 0. A significant rise in accuracy is seen for sums three and four which remains unexplained and requires further research. The data from the "Facebook Only" group indicate that such ties may be missed by most of a user's friends in the weakest stages, but perceived if that connection becomes marginally stronger.

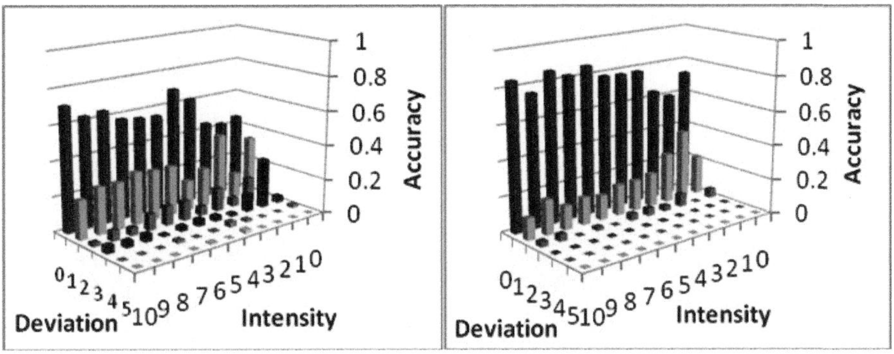

Fig. 8 (left) Compares perceived summed perceived connection intensity based on user self-reports between the participant and connection members on the x-axis to accuracy on the z-axis. The y-axis represents the closeness of a user response with 0 being exactly at baseline. User self-reports are excluded except to provide baseline. **(right).** Compares same parameters using others' perceptions as baseline. More accurate strength estimates are found under this second criterion. More research would be necessary to determine if this is an artifact of the selection algorithm. (n = 518 ratings)

This section has confirmed hypothesis (3), increased user closeness correlates with improved perceptual accuracy about social networks. It has also proven that others' perceptions highly correlate with self-reported connections, highlighting our high performance at memorizing these connections.

4.4 Inter/Intra Clique Comparison

Cliques are clusters of interconnected groups where every user is connected, defined as a completely connected subgraph. While complete data on the friendship networks of every user were unavailable, the sampling enacted yielded a significant fraction of users' friend networks as many overlapped significantly.

Figure 9 confirms Hypothesis(4), and indicates that more research may push it further, as it indicates that clique accuracy may be inversely proportionate to clique size, approaching the guessing rate, 0.5, as clique size increases. Inter-clique accuracy lies below the guessing rate, indicating that participants fell prey to Janicik and Larrick's transitivity problem in that they assumed offline connections were indicative of online connection, as many more errors were false positives than false negatives. It may also result from Facebook's incomplete map of users' social networks, or users' desire to not share online information with a subset of their friend group.

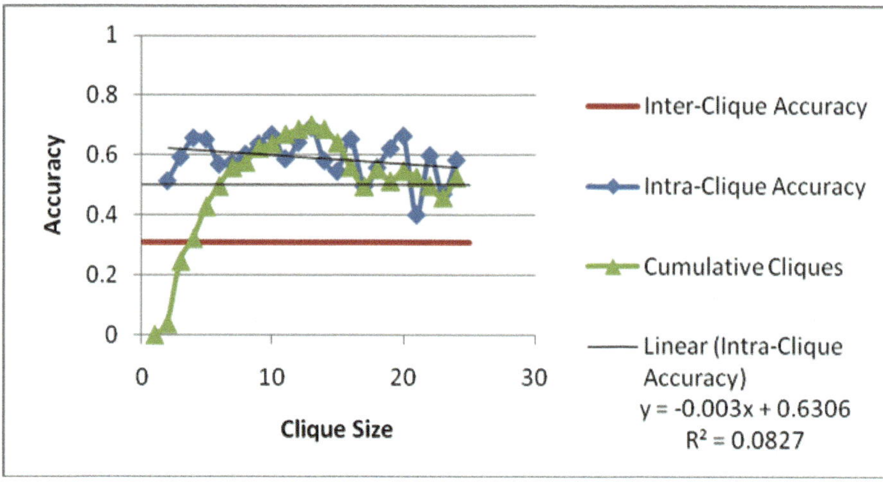

Fig. 9 Compares clique size (x-axis) with accuracy (y-axis). The full list of cliques is excerpted due to quantity detected (60,000 unique cliques). Inter-clique accuracy represents the accuracy of perceptions that bridge two cliques, found by examining edges not present in any found clique. Intra-clique accuracy represents accuracy in the maximal size clique found to contain the given connection. Cumulative cliques represents the average accuracy on all cliques of a given size. (n = 14,051 ratings)

5 Discussion

The results presented in this paper apply to many fields, including epidemiology, information transmission, network analysis, human behavior, economics, and neuroscience.

Constructing proper epidemiological models requires understanding both the structure of people's friend networks and their perceptions of the intensity among that network. Understanding the network's structure enables accurate modeling of disease transmission, as more tightly connected individuals tend to spend more time together both online and offline. Perceptions about others' connections are crucial in studies of sexually transmitted diseases as partners gauge the sexual histories of their mates. Online connections are especially interesting because in some cases they may supplant regular offline contact because of geographic separation, but will periodically interact in person, potentially transmitting any type of disease stochastically. These studies indicate that online connections may be an excellent place to look for connections that are not present in one's daily life, but must be modeled to account for disease transmission.

This data can also help model information flows online. People who are closer together will tend to share more data. That closeness will also mean that users will know who to ask to gain access to specific information. Being able to predict the best places to introduce information for rapid transmission to the entire network would be important for mass movements and responses to authoritarian repression, or the key nodes for information flow within a terrorist network.

This data also addresses a primary concern of social networking research by providing a ground truth to compare other connection annotation algorithms against. Given an algorithm, this data could be used to judge that algorithm's effectiveness at reproducing user reports regarding the intensity of their connections to others. If an effective algorithm were found, the graph could be annotated automatically, providing significant insight in the examples cited above.

Understanding how information enters the human brain is crucial to comprehending how we act. Fields like economics assume that we are actors who have access to a complete understanding of the world and all its information and act on that information rationally. This research gives insight into how online connections increase the information flowing to us, how we filter the information presented to us, and how we select which information to pay attention to. Much research has suggested that we respond to suggestions from our close friends very positively. By understanding when, where, and why we interact with those friends, we can better understand ourselves, and better refine Adam Smith's "rational actor."

This research highlights the capabilities of the human brain with respect to understanding social connections, indicating a flexible and accurate system for understanding those networks. Introducing the online world appears to enhance the quantity of these connections without reducing our perceptual accuracy, a crucial result that further supports hypotheses placing our understanding of those around us as a key element of what cognitively makes us human.

References

1. Boyd, D.M., Ellison, N.B.: Social network sites: Definition, history, and scholarship. Journal of Computer-Mediated Communication 13, 11 (2008)
2. Golder, S.A., Wilkinson, D., Huberman, B.A.: Rhythms of social interaction: Messaging within a massive online network. In: Steinfield, C., et al. (eds.) Proceedings of Third International Conference on Communities and Technologies, pp. 41–66. Springer, London (2007)
3. Feld, S.: The Focused Organization of Social Ties. The American Journal of Sociology 86, 1015–1035 (1981)
4. Lewis, K., et al.: Tastes, ties, and time: A new social network dataset using Facebook.com. Social Networks 30, 330–342 (2008)
5. Ellison, N., Steinfield, C., Lampe, C.: The benefits of Facebook 'friends': Exploring the relationship between college students' use of online social networks and social capital. Journal of Computer-Mediated Communication 12, 1 (2007)
6. Stone, B.: Is Facebook Growing Up Too Fast?" The New York Times March 29 (2009), http://www.nytimes.com/2009/03/29/technology/internet/29face.html?ref=technology
7. Wellman, B., Haase, A., Witte, J., Hampton, K.: Does the Internet Increase, Decrease, or Supplement Social Capital? American Behavioral Scientist 45, 436–455 (2001)
8. Donath, J., Boyd, D.M.: Public displays of connection. BT Technology Journal 22(4), 71–82 (2004)

9. Primates on Facebook. The Economist, http://www.economist.com/science/displaystory.cfm?story_id=13176775
10. Granovetter, M.: The strength of weak ties. American Journal of Sociology 78, 360–380 (1973)
11. Freeman, L.C., Freeman, S.C., Michaelson, A.G.: On human social intelligence. Journal of Social and Biological Structures 11, 415–425 (1988)
12. Janicik, G.A., Larrick, R.P.: Social network schemas and the learning of incomplete networks. Journal of Personality and Social Psychology 88, 348–364 (2005)

Ranking Learning Entities on the Web by Integrating Network-Based Features

Yingzi Jin, Yutaka Matsuo, and Mitsuru Ishizuka

Abstract. Many efforts are undertaken by people and companies to improve their popularity, growth, and power, the outcomes of which are all expressed as rankings (designated as target rankings). Are these rankings merely the results of those person's or that company's own attributes? In the theory of social network analysis (SNA), the performance and power of actors are usually interpreted as relations and the relational structures in which they are embedded. We propose an algorithm to generate and integrate network-based features systematically from a given social network that is mined from the world-wide web. After learning a model for explaining target rankings researchers' productivity based on social networks confirms the effectiveness of our models. This chapter specifically examines the application of a social network that exemplifies the advanced use of social networks mined from the web.

1 Introduction

People prefer to use rankings to compare companies, to discuss elections, and to evaluate goods. For example, investors seek to invest their funds in fast-growing and stable companies; consumers tend to buy highly popular products. Therefore, many

---

Yingzi Jin*
The University of Tokyo, IBM T.J. Watson Research Center, 19 Skyline Dr., Hawthorne, NY, USA
e-mail: yjin@us.ibm.com
*Research Fellow of the Japan Society for the Promotion of Science (JSPS)

Yutaka Matsuo
The University of Tokyo, 2–11–16 Yayoi, Bunkyou-ku, Tokyo, Japan
e-mail: matsuo@biz-model.t.u-tokyo.ac.jp

Mitsuru Ishizuka
The University of Tokyo, Hongo 7–3–1, Tokyo 113-8656, Japan
e-mail: ishizuka@i.u-tokyo.ac.jp

I.-H. Ting et al. (Eds.): Mining and Analyzing Social Networks, SCI 288, pp. 107–123.
springerlink.com

efforts have been undertaken by people and companies to improve their popularity, growth, and power, the outcomes of which are all expressed as rankings. Conventionally, these rankings are evaluated and ranked by values from statistical data and attributes of actors such as income, education, personality, and social status.

In the theory of social network analysis (SNA), social networks are used to analyze the performance and valuation of social actors [13]. Network researchers have argued that relational and structural embeddedness influence individuals' behavior and performance, and that a successful person (or company) must therefore emphasize relation management. Actually, several relations exist in the world with different impacts; the actors might be tied together closely in one relational network, but can differ greatly from one to another in a different relational network. The question therefore arises: *Relations of what kind are important for entities?* Unfortunately, the answers of important relations have been decided according to the judgments of researchers themselves.

To identify the prominence or importance of an individual actor embedded in a network, centrality measures have been used in social sciences: degree centrality, betweenness centrality, and closeness centrality. These measures often engender distinct results with different perspectives of "actor location", i.e. local (e.g. degree) and global (e.g. eigenvector) locations, in a social network [13]. Another question arises: *What kind of centrality indices are most appropriate for ranking actors?* That question can be extended as *What kind of structural embeddedness of actors makes them more powerful?*

This chapter presents a description of an attempt to learn the ranking of named entities from a social network that has been mined from the world-wide web. It enables us to have a model to rank entities for various purposes: one might wish to rank entities for search and recommendation, or might want to have the ranking model for prediction. Given a list of entities, we first extract relations of different types from the web based on our previous work [4, 8]. Subsequently, we rank the entities on these networks using different network indices. In this chapter, we propose a systematic algorithm that integrates features generated from networks (designated as *network-based features*) for each and then use these features to learn and predict rankings. We conducted experiments related to social networks among researchers to learn and predict the ranking of researchers' productivity.

The contributions of this study can be summarized as follows. We provide an example of advanced utilization of a social network mined from the web. The results illustrate the usefulness of our approach, by which we can understand the important relations as well as the important structural embeddedness to predict ranking of entities. The model can be combined with a conventional attribute-based approach. Results of this study will provide a bridge between relation extraction and rank learning to facilitate advanced knowledge (web intelligence) acquisition.

The following section presents a description of an overview of the ranking learning model. Section 3 briefly introduces our previous work for extracting social networks from the web. Section 4 describes the proposed ranking learning models based on extracted social networks. Section 5 explains the experimental settings and results. Section 6 presents some related works before the chapter concludes.

2 System Overview

Our study explores the integration of mining relations (and structures) among entities and the learning ranking of entities. For that reason, we first extract relations and then determine a model based on those relations. Our reasoning is that important relations can be recognized only when we define some tasks. These tasks include ranking or scores for entities, i.e. *target ranking*, such as ranking of companies, CD sales, popular blogs, and sales of products. In short, our approach consists of two steps:

Step 1: Constructing Social Networks. Given a list of entities with a target ranking, we extract a set of social networks among these entities from the web.
Step 2: Ranking learning. Learn a ranking model based on relations and structural features generated from the networks.

Once we obtain a ranking model, we can use it for prediction for unknown entities. Additionally, we can obtain the weights for each relation type as well as the relation structure, which can be considered as important for target rankings. The social network can be visualized by specifically examining its inherent relations if the important relations are identified. Alternatively, social network analysis can be executed based on the relations.

3 Constructing Social Networks

In this step, our task is, given a list of entities $V = \{v_1, \dots, v_n\}$, to construct a set of social networks $G_i(V, E_i)$, $i \in \{1, \dots, m\}$ where m signifies the number of relations, and $E_i = \{e_i(v_x, v_y) | v_x \in V, v_y \in V, v_x \neq v_y\}$ denotes a set of edges with respect to the i-th relation.

A social network is obtainable through various approaches [4, 8, 9]. In this chapter, we detail the web mining approaches—*co-occurrence-based approach* and *classification-based approach*—as a basis of our study. For the *co-occurrence-based approach* [8, 9], given a person name list, the strength of relevance of two persons, x and y, is estimated by putting a query x AND y to a search engine. An edge will be invented when the relation strength by the co-occurrence measure is higher than a predefined threshold. Subsequently, we extract co-occurrence-based networks of two kinds: cooc network (G_{cooc}), and overlap network ($G_{overlap}$). The relational indices are calculated respectively using the matching coefficient $n_{x \wedge y}$ and the overlap coefficient $n_{x \wedge y} / \min(n_x, n_y)$, where n_k means the number of hits obtained after issuing query k to a search engine. For the *classification-based approach* [8] based on web co-occurrence networks, edges are classified into those representing one of several relations using C4.5 as a classifier. In our experiments, we first extract overlap network among researchers, then classify the edges into relational networks of two kinds: an co-affiliation network ($G_{affiliation}$) and a co-project network ($G_{project}$). Because of space limitations, we show no details related to the construction algorithms. Details are provided in an earlier report [8]. Extracted networks for 253 researchers are portrayed in Fig. 1. It is apparent that social networks vary

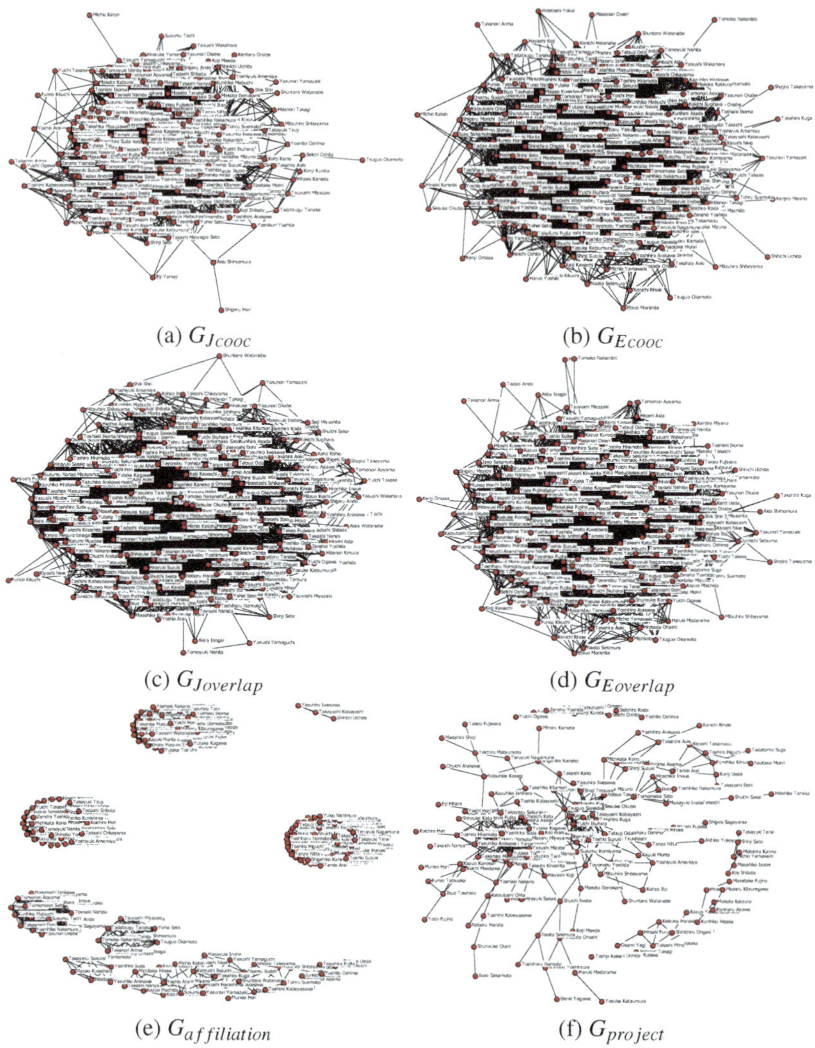

(a) G_{Jcooc} (b) G_{Ecooc}

(c) $G_{Joverlap}$ (d) $G_{Eoverlap}$

(e) $G_{affiliation}$ (f) $G_{project}$

Fig. 1 Web-based social networks for researchers with different relational indices or types

with different relational indices or types even though they contain the same list of entities.

4 Ranking Learning Model

For the list of nodes $V = \{v_1, \ldots, v_n\}$, given a set of networks $G_i(V, E_i)$, $i \in \{1, \ldots, m\}$ (constructed by section 3) with a target ranking $\mathbf{r}^*$ ($\in R^t$) (where $t \leq n$, and r_k^* denotes k-th element of the vector $\mathbf{r}^*$ and means the target ranking score of entity v_k), the goal is to learn a ranking model based on these networks.

First, as a baseline approach, we follow the intuitive idea of simply using the approach from SNA (i.e. centrality) to learn ranking. Then we propose a more systematic algorithm that generates various network features for individuals from social networks.

4.1 Baseline Model

Based on the intuitive approach, we first overview commonly used indices in social network analysis and complex network studies. Given a set of social networks, we rank entities on these networks using different network centrality indices. We designate these rankings as *network rankings* because they are calculated directly from relational networks.

To address the question of what kind of relation is most important for entities, we intuitively compare rankings resulting from relations of various types. Although simple, it can be considered as an implicit step of social network analysis given a set of relational networks. We merely choose the type of relation that maximally explains the given ranking. We rank the relational network of each type; then we compare the *network ranking* with the *target ranking*. Intuitively, if the correlation to the network ranking $\mathbf{r}_{\hat{i}}$ is high, then the relation $\hat{i}$ represents important influences among entities for the given target ranking. Therefore, this model is designed to determine an optimal relation $\hat{i}$ from a set of relations:

$$\hat{i} = \underset{i \in \{1,\ldots,m\}}{\operatorname{argmax}} \, Cor(\mathbf{r}_i, \mathbf{r}^*) \,, \tag{1}$$

For different relational networks with different centrality indices, the network ranking from i-th network with j-th centrality ranking can be represented as $\mathbf{r}_{i,j}$ ($\in R^n$), where $i \in \{1,\ldots,m\}$, and $j \in \{1,\ldots,s\}$. Therefore, the first method can be extended simply to find a pair of optimal parameters $< \hat{i}, \hat{j} >$ (i.e. the i-th network by j-th centrality rankings) that maximizes the coefficient between network rankings with a target ranking.

$$< \hat{i}, \hat{j} > = \underset{i \in \{1,\ldots,m\} \; j \in \{1,\ldots,s\}}{\operatorname{argmax}} \; Cor(\mathbf{r}_{i,j}, \mathbf{r}^*) \,, \tag{2}$$

4.2 Network Combination Model

Many centrality approaches related to ranking network entities specifically examine graphs with a single link type. However, multiple social networks exist in the real world, each representing a particular relation type; each of which might be integrated to play a distinct role in a particular task. We combine several extracted multiple social networks into one network and designate such a social network as a *combined-relational network* (denoted as $G_c(V, E_c)$). Our target is using a combined-relational network—which is integrated with multiple networks extracted from the web—to learn and predict the ranking. The important question that must be resolved here is *how to combine relations to describe a given ranking best.*

For $G_c(V,E_c)$, the set of edges is $E_c = \{e_c(v_x,v_y)|v_x \in V, v_y \in V, v_x \neq v_y\}$. Using a linear combination, each edge $e_c(v_x,v_y)$ can be generated from $\sum_{i\in\{1,\ldots,m\}} w_i e_i(v_x,v_y)$, where w_i is the i-th element of $\mathbf{w}$ (i.e. $\mathbf{w} = [w_1,\ldots,w_m]^T$). Therefore, the purpose is to learn optimal combination weights $\hat{\mathbf{w}}$ to combine relations as well as optimal ranking method h_j on G_c:

$$< \hat{\mathbf{w}}, \hat{j} >= \underset{\mathbf{w},h_j\in\{h_1,\ldots,h_s\}}{\operatorname{argmax}} \; Cor(\mathbf{r}_{c,j},\mathbf{r}^*). \tag{3}$$

Cai et al. [3] examine a similar idea with this approach: They attempt to identify the best combination of relations (i.e. relations as features) which makes the relation between the intra-community examples as tight as possible. Simultaneously, the relation between the inter-community examples is as loose as possible when a user provides multiple community examples (e.g. two groups of researchers). However, our purpose is learn a ranking model (e.g. ranking of companies) based on social networks, which has a different optimization task. Moreover, we propose innovative features for entities based on combination or integration of structural importance generated from social networks.

For this study, we simply use Boolean type ($w_i \in \{1,0\}$) to combine relations. Using relations of m types to combine a network, we can create $2^m - 1$ types of combination-relational networks (in which at least one type of relation exists in the G_c). We obtain network rankings in these combined networks to learn and predict the target rankings. Future work on how to choose parameter values will be helpful to practitioners.

4.3 Network-Based Feature Integration Model

The proposed method in our research is to integrate multiple indices that are obtained from multiple social networks to learn the target rankings. A feature by itself (e.g. a centrality value) might have little correlation with the target ranking, but when it is combined with some other features, they might be strongly correlated with the target rankings [14]. Simply, we can integrate various centrality values for each actor, thereby combining different meanings of importance to learn the ranking. Furthermore, we can generate additional relational and structural features from a network for each, such as how many nodes are reachable, how many connections one's friends have, and the connection status of one's friends. We might understand something about the behavior and power of the individual while we predict their ranking if we could know the structural position of individuals. Herein, we designate these features generated from networks as *network-based features*. The interesting question is *how to generate network-based features from networks for each*, and *how to integrate these features to learn and predict rankings*. Below we will describe the approach.

4.3.1 Generating Network-Based Features for Nodes

For each x, we first define node sets with relations that might affect x. We define a set of nodes $C_x^{(k)}$ as a set of nodes within distance k from x. We choose a node set adjacent to node x (designated as C_x^1), and also choose a node set that contains all reachable nodes from x (designated as $C_x^{(\infty)}$) as influential nodes for x.

Then we apply some operators to the set of nodes to produce a list of values. The simple operation for two nodes is to check whether the two nodes are adjacent or not. We denote these operators as $s^{(1)}(x,y)$, which returns 1 if nodes x and y are mutually connected, and 0 otherwise. We also define operator $t(x,y) = argmin_k\{s^{(k)}(x,y) = 1\}$ to measure the geodesic distance between the two nodes on the graph. These two operations are applied to each pair of nodes in nodeset N, which is definable as $Operator \circ N = \{Operator(x,y) | x \in N, y \in N, x \neq y\}$. For example, if we are given a node set $\{ n_1, n_2, n_3\}$, then we can calculate $s^{(1)}(n_1,n_2)$, $s^{(1)}(n_1,n_3)$, and $s^{(1)}(n_2,n_3)$ and return a list of three values, e.g. $(1,0,1)$. We denote this operation as $s^{(1)} \circ N$. In addition to s and t operations, we define two other operations. One operation is to measure the distance from node x to each node, denoted as t_x. Instead of measuring the distance between two nodes, $t_x \circ N$ measures the distance of each node in N from node x. Another operation is to check the shortest path between two nodes. Operator $u_x(y,z)$ returns 1 if the shortest path between y and z includes node x. Consequently, $u_x \circ N$ returns a set of values for each pair of $y \in N$ and $z \in N$.

Subsequently, the values calculated using the operations explained above are aggregated into a single feature value. Given a list of values, we can take the summation (*Sum*), average (*Avg*), maximum (*Max*), and minimum (*Min*). For example, if we apply *Sum* aggregation to a value list $(1,0,1)$, then we obtain a value of 2. We can write the aggregation as, for example, $Sum \circ s^{(1)} \circ N$. Although other operations can be performed, such as taking the variance or taking the mean, we limit the operations to the four described above. The value obtained here results in the network-based feature for node x. Additionally, we can take the difference or the ratio of two obtained values. For example, if we obtain 2 by $Sum \circ s^{(1)} \circ C_x^{(1)}$ and 1 by $Sum \circ s^{(1)} \circ C_x^{(k)}$, then the ratio is 2/1 = 2.0.

The nodesets, operators, and aggregations are presented in Table 1. We have 2 (nodesets) × 5 (operators) × 4 (aggregations) = 40 combinations. There are ratios for $C_x^{(1)}$ to $C_x^{(k)}$ if we consider the ratio. In all, 4 × 5 more combinations also exist: there are 60 in all. Each combination corresponds to a feature of node x. The resultant value sometimes corresponds to a well-known index, as we had intended in the design of the operators. For example, the degree centrality can be expressed as $Sum \circ s_x^{(1)} \circ C_x^{(1)}$, and closeness centrality is expressed as $Avg \circ t_x \circ C_x^{(\infty)}$. These features represent some possible combinations. Some lesser-known features might actually be effective.

4.3.2 Network-Based Features with SNAs Indices

It is readily apparent that centralities described in the baseline approach are also a particular case of this model because our network-based features include those

Table 1 Operator list

Notation	Input	Output	Description
$C_x^{(1)}$	node x	a nodeset	adjacent nodes to x
$C_x^{(k)}$	node x	a nodeset	nodes within distance k from x
$s^{(1)}$	a nodeset	a list of values	1 if connected, 0 otherwise
t	a nodeset	a list of values	distance between a pair of nodes
t_x	a nodeset	a list of values	distance between node x and other nodes
γ	a nodeset	a list of values	number of links in each node
u_x	a nodeset	a list of values	1 if the shortest path includes node x, 0 otherwise
Avg	a list of values	a value	average of values
Sum	a list of values	a value	summation of values
Min	a list of values	a value	minimum of values
Max	a list of values	a value	maximum of values
$Ratio$	two values	value	ratio of value on neighbor nodeset $C_x^{(1)}$ by reachable nodeset $C_x^{(\infty)}$

centrality measures and other SNAs indices for each node. Below, we describe other examples used in the social network analysis literature.

- network diameter: $Min \circ t \circ N$
- characteristic path length: $Avg \circ t \circ N$
- degree centrality: $Sum \circ s_x^{(1)} \circ C_x^{(1)}$
- node clustering: $Avg \circ s^{(1)} \circ C_x^{(1)}$
- closeness centrality: $Avg \circ t_x \circ C_x^{(\infty)}$
- betweenness centrality: $Sum \circ u_x \circ C_x^{(\infty)}$,
- structural holes: $Avg \circ t \circ C_x^{(1)}$

When we set the element $Sum \circ s_x^{(1)} \circ N_x^{(1)}$ in a feature vector equal to 1, and all others to 0, we can elucidate the effect of degree centrality for predicting target ranking.

4.3.3 Network-Based Feature Integration

After we generate various network-based features for individual nodes, we integrate them to learn the ranking. We introduce an f-dimensional feature vector F, in which each element represents a network-based feature for each node. We identify the f-dimensional combination vector $\mathbf{u} = [u_1, \ldots, u_f]^T$ to combine network-based features for each node. The inter-product $\mathbf{u}^T \mathbf{F}$ for each node produces an n-dimensional ranking. For relational networks of m kinds, the feature vector can be expanded to $m \times 60$ dimensions. In this case, the purpose is finding out whether optimal combination weight $\hat{\mathbf{u}}$ maximally explains the target ranking:

$$\hat{\mathbf{u}} = \underset{\mathbf{u}}{\operatorname{argmax}} Cor(\mathbf{u}^T \bullet F, \quad \mathbf{r}^*) , \tag{4}$$

This model can be augmented easily with other traditional attributes of entities as features. We can use any technique such as SVM, boosting, and neural networks to implement the optimization problem. For multi-relational networks, we can generate features for each single-relational network. Thereby, we can compare the performance among them to elucidate which relational network produces more reasonable features. We can determine which relation(s) is important for the target ranking.

5 Experimental Results

In this section, we describe results to clarify the effectiveness of ranking learning on extracted social networks. We use data of 253 researchers from The University of Tokyo to predict a ranking of researchers. In our experiments, we conducted three-fold cross-validation. In each trial, two folds of actors are used for training, and one fold for prediction. The results we report in this section are those averaged over three trials. We use Spearman's rank correlation coefficient to measure the pairwise ranking correlation between predicted rankings and the target ranking.

5.1 Datasets

We extract social networks for researchers (253 professors of The University of Tokyo) to learn and predict the ranking of researchers. We use the ranking by the number of publications (designated as Paper) as a target ranking, as presented in Table 2. Academic papers are often the product of several researchers' collaboration. Therefore, a good position in a social network is derived through good performance. Is there any relation that is important to predict productivity?

We construct social networks among researchers from the web using a general search engine. We detail the co-occurrence-based approach (Section 6.3.1) to extract co-occurrence-based networks of two kinds in English-language web sites and Japanese web sites respectively: a cooc network (G_{Ecooc}, G_{Jcooc}) and an overlap network ($G_{Eoverlap}$, $G_{Joverlap}$). Actually, we used English/romanized names of researchers as a query to obtain co-occurrence information for G_{Ecooc} and $G_{Eoverlap}$, and used Japanese names of researchers as a query to obtain co-occurrence information for G_{Jcooc} and $G_{Joverlap}$. Then, based on web co-occurrence networks (using Japanese web sites), we use the context of web pages retrieved using two names of persons to classify the relations using C4.5 as a classifier (details presented in [8]). We use a Jaccard network constructed using the approach described above; then we classify the edges into relational networks of two kinds: a co-affiliation network ($G_{affiliation}$) and a co-project network ($G_{project}$). Extracted networks for 253 researchers are portrayed in Fig. 1.

For this experiment, we also use researcher attributes of two types: the number of hits on Japanese web sites JhitNum (using Japanese names as a query) and the number of hits on the English-language web sites EhitNum) (using English/ romanized names as a query).

Table 2 Ranking of the number of pages for the top 50 researchers of The University of Tokyo

r*	Name	**r***	Name
1:	Yasuhiko Arakawa	26:	Kazuhiko Saigo
2:	Kazunori Kataoka	27:	Tadatomo Suga
3:	Kohji Kishio	28:	Tamio Arai
4:	Yuichi Ikuhara	29:	Akira Isogai
5:	Kazuhiko Ishihara	30:	Ryoichi Yamamoto
6:	Yasuhiro Iwasawa	31:	Takayasu Sakurai
7:	Genki Yagawa	32:	Michio Yamawaki
8:	Kazuhito Hashimoto	33:	Hiroshi Harashima
9:	Hiroyuki Sakaki	34:	Takayoshi Kobayashi
10:	Hideki Imai	35:	Fumio Tatsuoka
11:	Masaharu Oshima	36:	Takehiko Kitamori
12:	Kazuyuki Aihara	37:	Teruyuki Nagamune
13:	Kazuro Kikuchi	38:	Masahiko Isobe
14:	Yoshiaki Nakano	39:	Motohiro Kanno
15:	Shinichi Uchida	40:	Kazuo Hotate
16:	Hidenori Takagi	41:	Mitsuhiro Shibayama
17:	Hiroyuki Fujita	42:	Hajime Asama
18:	Katsushi Ikeuchi	43:	Satoru Tanaka
19:	Yutaka Kagawa	44:	Isao Shimoyama
20:	Nobuo Takeda	45:	Yozo Fujino
21:	Masaru Miyayama	46:	Takayuki Terai
22:	Toshiro Higuchi	47:	Yoichiro Matsumoto
23:	Tsuguo Sawada	48:	Nobuhide Kasagi
24:	Kiyoharu Aizawa	49:	Yoshiyuki Amemiya
25:	Kimihiko Hirao	50:	Kunihiro Asada

In our experiments, we conducted three-fold cross-validation. In each trial, two folds of actors are used for training, and one fold for prediction. The results reported in this section are those averaged over three trials. We use Spearman's rank correlation coefficient (ρ) [11] to measure the pairwise ranking correlation.

$$\rho = 1 - \frac{6\Sigma_i^2}{n(n^2-1)} \tag{5}$$

In that equation, d_i signifies the difference between the ranks of corresponding values X_i and Y_i.

5.2 Ranking Results

First, we rank researchers on different network rankings. Table 3 presents the degree centrality rankings of different types of networks in researcher networks. Results show that *Yutaka Kagawa* has good degree centrality on a cooc network of Japanese

Table 3 Top 20 researchers ranked by degree centrality on different social networks

$\mathbf{r}_{i,Cd}$	$\mathbf{r}_{Ecooc,Cd}$	$\mathbf{r}_{Eoverlap,Cd}$	$\mathbf{r}_{Jcooc,Cd}$
1:	Masatoshi Ishikawa	Yoshiaki Nakano	Yutaka Kagawa
2:	Yasuhiko Arakawa	Hiroyuki Fujita	Masatoshi Ishikawa
3:	Masaru Kitsuregawa	Susumu Tachi	Masaru Kitsuregawa
4:	Genki Yagawa	Akira Watanabe	Yasuhiko Arakawa
5:	Yutaka Kagawa	Masataka Fujino	Tsuguo Sawada
6:	Yasuhiro Iwasawa	Koji Maeda	Yasuhiro Iwasawa
7:	Masao Kuwahara	Kunihiko Mabuchi	Keiji Kawachi
8:	Kiyoharu Aizawa	Yasushi Mizobe	Makoto Kuwabara
9:	Takahisa Masuzawa	Isao Shimoyama	Genki Yagawa
10:	Toshimi Kabeyasawa	Kazuro Kikuchi	Masao Kuwahara
11:	Koichi Maekawa	Taketo Uomoto	Kazuhiko Hirakawa
12:	Takeo Fujiwara	Takeshi Kinoshita	Takahisa Masuzawa
13:	Yuichi Ogawa	Yoichi Hori	Masanori Owari
14:	Shuichi Iwata	Tamaki Ura	Takeo Fujiwara
15:	Makoto Kuwabara	Kazuyuki Aihara	Kiyoharu Aizawa
16:	Tsuguo Sawada	Chisachi Kato	Chuichi Arakawa
17:	Kazuhiko Hirakawa	Kenshiro Takagi	Shuichi Iwata
18:	Ikuo Towhata	Kohji Kishio	Koichi Maekawa
19:	Chuichi Arakawa	Takayasu Sakurai	Ikuo Towhata
20:	Yoshihiro Arakawa	Hideyuki Suzuki	Hitoshi Kuwamura
$\mathbf{r}_{i,Cd}$	$\mathbf{r}_{Joverlap,Cd}$	$\mathbf{r}_{affiliation,Cd}$	$\mathbf{r}_{project,Cd}$
1:	Shoji Tetsuya	Koichi Maekawa	Yutaka Kagawa
2:	Haruo Yoshiki	Michio Yamawaki	Kazuhiko Hirakawa
3:	Yasuhiro Tani	Keiji Kawachi	Tsuguo Sawada
4:	Shigefumi Nishio	Ikuo Towhata	Masanori Owari
5:	Michikata Kono	Genki Yagawa	Masao Kuwahara
6:	Seisuke Okubo	Hitoshi Kuwamura	Yasuhiko Arakawa
7:	Michio Katoh	Yoshihiro Arakawa	Makoto Kuwabara
8:	Shigehiko Kaneko	Shuichi Iwata	Takahisa Masuzawa
9:	Akio Shimomura	Makoto Kuwabara	Koji Araki
10:	Koji Araki	Takeo Fujiwara	Hidetoshi Yokoi
11:	Minoru Kamata	Takahisa Masuzawa	Shuichi Iwata
12:	Hideaki Miyata	Kazuhiko Hirakawa	Jun Yanagimoto
13:	Tomoko Nakanishi	Yutaka Kagawa	Yasushi Mizobe
14:	Hiroshi Hosaka	Masaru Kitsuregawa	Ikuo Towhata
15:	Hitoshi Kuwamura	Kiyoharu Aizawa	Taketo Uomoto
16:	Eiji Hihara	Tsuguo Sawada	Koichi Maekawa
17:	Yutaka Toi	Masatoshi Ishikawa	Kenichi Hatanaka
18:	Yutaka Kagawa	Takayuki Terai	Susumu Nanao
19:	Tomonari Yashiro	Shigeru Morichi	Yasuhiro Iwasawa
20:	Kenichi Hatanaka	Noritaka Mizuno	Yoshihiro Arakawa

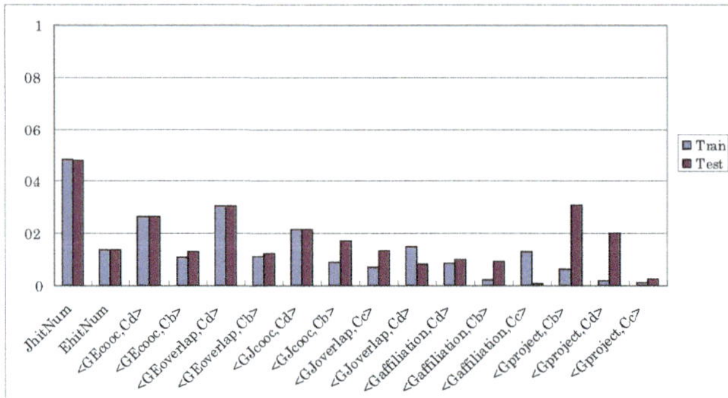

Fig. 2 Evaluation for each attribute-based ranking as well as centrality-based ranking with target ranking among researchers

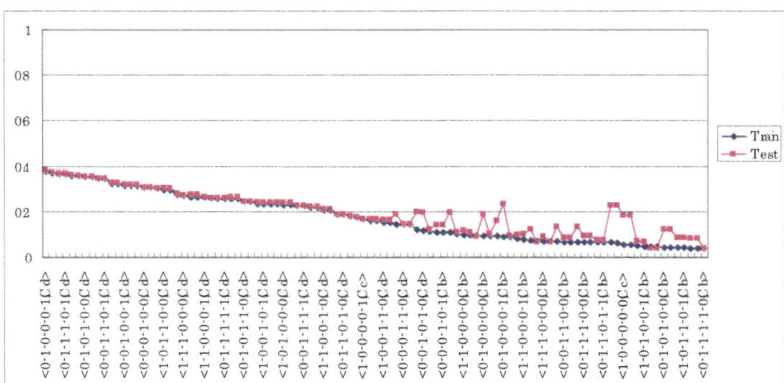

Fig. 3 Evaluation for network rankings in a combined-relational network with Paper among researchers

web sites G_{Jcooc} and a co-affiliation network $G_{affiliation}$, and *Masaru Kitsuregawa* has stable centralities on several networks.

For the baseline model, three centrality indices (degree centrality C_d, closeness centrality C_c, and betweenness centrality C_b) are used on different networks (G_{Ecooc}, $G_{Eoverlap}$, G_{Jcooc}, $G_{Joverlap}$, $G_{affiliation}$, and $G_{project}$) as network rankings. We calculate the correlation between network rankings with each target ranking of Paper. For comparison, we also rank companies according to previously described attributes (i.e., JhitNum and EhitNum), and take correlation with target ranking. Actually, Fig. 2 portrays correlations (mean of three tries) of each network ranking as well as each attribute-based ranking with different target rankings on training and testing data among researchers. Results show that the hit number of names on Japanese web sites is a good attribute of researchers for predicting the creditability of publications. Furthermore, degree centralities in an overlap network, as they do in a cooc network on English-language web sites ($\mathbf{r}_{G_{Eoverlap},C_d}$ and $\mathbf{r}_{G_{Ecooc},C_d}$) exhibit

Table 4 Results of feature integration among researchers

Professor	Feature	PaperNum Train	PaperNum Test
Network	G_{Ecooc}	0.470	0.413
	$G_{Eoverlap}$	0.508	0.411
	G_{Jcooc}	0.443	0.261
	$G_{Joverlap}$	0.585	0.325
	$G_{affiliation}$	0.178	-0.011
	$G_{project}$	0.540	0.043
	G_{ALL}	0.821	**0.417**
Attributes	ALL	0.491	**0.448**
Network + Attributes	G_{Ecooc}+A	0.514	0.429
	$G_{Eoverlap}$+A	0.544	0.404
	G_{Jcooc}+A	0.481	0.284
	$G_{Joverlap}$+A	0.519	0.420
	$G_{affiliation}$+A	0.497	0.159
	$G_{project}$+A	0.548	0.304
	G_{ALL}+A	0.811	**0.456**

a good correlation with target ranking. One might infer that researchers who are famous on Japanese web sites and who frequently co-occur with other researchers on English-language web sites are the more creative researchers.

In the combination model, we also use Boolean type ($w_i \in \{1,0\}$) operators to combine the relations. Using relations of six types to combine a network $G_{affiliation-Ecooc-Eoverlap-Jcooc-Joverlap-project}$, we can create $2^6 - 1$ (=63) types of combination-relational networks (in which at least one type of relation exists). We obtain network rankings in this combined network to learn and predict the target rankings. The top 50 correlations between network rankings in a combined-relational network and target rankings are portrayed in Fig. 3. Results show that degree centralities on combined-relational network produce good correlation with target rankings. For instance, combining cooc relations on English-language web sites with co-project relations ($G_{0-1-0-0-0-1}$), or combining a cooc relation and overlap relations on English-language web sites with a cooc relation on Japanese web sites ($G_{0-1-1-1-0-0}$) makes the networks more reasonable for use in predicting a target ranking.

We execute our feature integration ranking model (with several variations) to single and multiple relational social networks to train and predict rankings of researchers' Paper. We use Ranking SVM to learn the ranking model which minimizes the pairwise training error in the training data. Then we apply the model to predict rankings on training data (again) and on testing data. Table 4 presents comparable results for models of several types. First, we integrate attribute indices (i.e., hit number of names on the Japanese web sites and on English-language web sites) of researchers as features, thereby producing a baseline of this model to learn and predict the rankings. We can obtain a 0.448 correlation coefficient between predicted rankings and target rankings, which seems to be readily explainable: famous researchers are also famous on the web sites. Subsequently, we integrate the proposed network-based features obtained from each type of single network as well as multi-relational networks among researchers to train and predict the rankings. The co-occurrence-based networks G_{Ecooc}, $G_{Eoverlap}$, $G_{Joverlap}$ (especially on English-language web sites) appear to be a better explanation of target ranking of Paper than the co-affiliation network $G_{affiliation}$ or the co-project network $G_{projext}$. Using features from multi-relational networks G_{ALL}, the prediction results are better than for any other single-relational network. Furthermore, when we combine network-based features with attribute-based features to learn the model, the results outperform each using attribute-based features only and network-based features only.

5.3 Detailed Analysis of Useful Features

We use network-based features separately for training. We expect the target rankings to clarify their usefulness. Leaving out one feature, the others are used to train and predict the rankings to evaluate network-based features. In fact, the k-th feature is a useful feature for explaining the target ranking if the result worsens much when leaving out the k feature. Table 5 presents the effective features for the target ranking of Paper in the researcher field. For example, the maximum number of links in the reachable nodeset of x from the cooc network from English-language web sites $Max \circ \gamma \circ C_x^{(\infty)} \circ G_{Ecooc}$ is effective for the target ranking, which means that if a

Table 5 Effective features in various networks for Paper among researchers

Top	Effective Features for Paper
1	$Max \circ \gamma \circ C_x^{(\infty)} \circ G_{Ecooc}$
2	$Min \circ \gamma \circ C_x^{(1)} \circ G_{Jcooc}$
3	$Avg \circ \gamma \circ C_x^{(\infty)} \circ G_{Eoverlap}$
4	$Max \circ t \circ C_x^{(\infty)} \circ G_{Joverlap}$
5	$Avg \circ u_x \circ C_x^{(1)} \circ G_{Eoverlap}$
6	$Min \circ \gamma \circ C_x^{(1)} \circ G_{Eoverlap}$
7	$Min \circ \gamma \circ C_x^{(\infty)} \circ G_{Jcooc}$
8	$Ratio \circ (Sum \circ s^{(1)} \circ C_x^{(1)}, Sum \circ s^{(1)} \circ C_x^{(\infty)}) \circ G_{project}$
9	$Avg \circ \gamma \circ C_x^{(1)} \circ G_{Joverlap}$
10	$Min \circ \gamma \circ C_x^{(1)} \circ G_{Ecooc}$
11	$Ratio \circ (Sum \circ s^{(1)} \circ C_x^{(1)}, Sum \circ s^{(1)} \circ C_x^{(\infty)}) \circ G_{Ecooc}$
12	$Ratio \circ (Sum \circ u_x \circ C_x^{(1)}, Sum \circ u_x \circ C_x^{(\infty)}) \circ G_{Ecooc}$
13	$Min \circ u_x \circ C_x^{(1)} \circ G_{Jcooc}$
14	$Ratio \circ (Avg \circ u_x \circ C_x^{(1)}, Avg \circ u_x \circ C_x^{(\infty)}) \circ G_{Jcooc}$
15	$Min \circ \gamma \circ C_x^{(\infty)} \circ G_{Joverlap}$

famous researcher is reachable from a person, then that person can be more productive. The minimum number of links in the neighbor nodeset of x from the cooc network from Japanese web sites $Min \circ \gamma \circ C_x^{(1)} \circ G_{Jcooc}$ is also effective, which means that if a direct neighbor is productive, then x will be more productive. The ratio of the number of edges among neighbors to the number of edges among reachable nodes from co-project network $Ratio \circ (Sum \circ s^{(1)} \circ C_x^{(1)}, Sum \circ s^{(1)} \circ C_x^{(\infty)}) \circ G_{project}$ means that binding neighbors from all reachable nodes in projects makes the researcher more productive.

We understand that various features have been shown to be important for real-world rankings (i.e. target rankings). Some indices correspond to well-known indices in social network analysis: degree centrality, closeness centrality, and betweenness centrality. Some indices seem new, but their meanings resemble those of the existing indices. The results support the usefulness of the indices that are commonly used in the social network literature, and underscore the potential for additional composition of useful features.

Summary: Social networks vary according to different relational indices or types even if they contain the same list of researchers. Researchers have different centrality rankings even though they are in relational networks of the same type. Multi-relational networks have more information than single networks to explain target rankings. Well-chosen attribute-based features offer good performance for explaining target rankings. However, by combining proposed network-based features, the prediction results are further improved. Various network-based features have been shown to be important for real-world rankings (i.e. target rankings), some of which correspond to well-known indices in social network analysis such as degree centrality, closeness centrality, and betweenness centrality. Some indices seem new, but their meanings resemble those of existing indices.

6 Related Works

In the context of information retrieval, PageRank [10] and HITS [6] algorithms can be considered as well known examples for ranking web pages based on the link structure. Recently, algorithms that are more advanced have been proposed for learning to rank entities. Although numerous studies of learning-to-rank fields (particularly targeted on information retrieval) have investigated many attribute-based ranking functions learned from given preference orders, only a few studies have concluded that such an impact arises from relations and structures [1, 12]. Furthermore, our model is target-dependent: the important features of relations and structural embeddedness vary among different tasks.

Relations and structural embeddedness influence the behavior of individuals and growth and change of the group [13]. Several researchers use network-based features for analyses. Backstrom et al. [2] describe analyses of community evolution; they show some structural features characterizing individuals and positions in the network. Liben-Nowell et al. [7] elucidate features using network structures in the link prediction problem. We specifically examine relations and structural features for individuals (previously for link prediction in [5]) and address various structural features from multiple networks systematically for learning real world rankings (i.e. target rankings).

7 Conclusion

This paper described methods of learning the ranking of entities from social networks mined from the web. We first extracted social networks of different kinds from the web. Subsequently, we used these networks and a given target ranking to learn a ranking model. We proposed an algorithm to learn the model by integrating network-based features from a given social network that was mined from the web. We proposed three approaches used to obtain the ranking model. Results of experiments on a researcher field reveal the effectiveness of our models for explaining a target ranking of researchers' productivity using multiple social networks mined from the web. The results underscore the usefulness of our approach, with which we can elucidate important relations as well as important structural embeddedness to predict the rankings. Our model provides an example of advanced use of a social network mined from the web. More networks and attributes for various target rankings in different domains can be designated for improving the usefulness of our models in the future.

References

1. Agarwal, A., Chakrabarti, S., Aggarwal, S.: Learning to rank networked entities. In: The 12th ACM SIGKDD International Conference on Knowledge Discovery and Data mining, Philadelphia, USA (2006)
2. Backstrom, L., Huttenlocher, D., Lan, X., Kleinberg, J.: Group formation in large social networks: Membership, Growth, and Evolution. In: 12th ACM SIGKDD International Conference on Knowledge Discovery and Data mining, Philadelphia, USA (2006)
3. Cai, D., Shao, Z., He, X., Yan, X., Han, J.: Mining Hidden Community in Heterogeneous Social Networks. In: Proceedings of the ACM Workshop on Link Analysis and Group Detection, Chicago, USA (2005)
4. Jin, Y., Matsuo, Y., Ishizuka, M.: Extracting Social Networks Among Various Entities on the Web. In: Franconi, E., Kifer, M., May, W. (eds.) ESWC 2007. LNCS, vol. 4519, pp. 251–266. Springer, Heidelberg (2007)
5. Karamon, J., Matsuo, Y., Ishizuka, M.: Generating Useful Network-based Features for Analyzing Social Network. In: 23rd Conference on Artificial Intelligence, Chicago, USA (2008)
6. Jon, K.M.: Authoritative Sources in a Hyperlinked Environment. In: Proceedings of ACM-SIAM Symposium on Discrete Algorithms, pp. 668–677 (1998)
7. Liben-Nowell, D., Kleinberg, J.: The link prediction problem for social networks. In: 12th International Conference on Information and Knowledge Management, New Orleans, LA, USA (2003)
8. Matsuo, Y., Mori, J., Hamasaki, M., Ishida, K., Nishimura, T., Takeda, H., Hasida, K., Ishizuka, M.: POLYPHONET: an advanced social network extraction system. In: 15th International World Wide Web Conference, Edinburgh, Scotland (2006)

9. Mika, P.: Flink: semantic web technology for the extraction and analysis of social networks. Journal of Web Semantics 3(2), 211–223 (2005)
10. Page, L., Brin, S., Motwani, R., Winograd, T.: The Page Rank Citation Ranking: Bringing order to the Web. Stanford University (1998)
11. Spearman, C.: The proof and measurement of association between two things. Amer. J. Psychol. 15, 72–101 (1904)
12. Qin, T., Liu, T., Zhang, X., Wang, D., Xiong, W., Li, H.: Learning to Rank Relational Objects and Its Application to Web Search. In: 17th International World Wide Web Conference, Beijing, China (2008)
13. Wasserman, S., Faust, K.: Social network analysis, methods and applications. Cambridge University Press, Cambridge (1994)
14. Zhao, Z., Liu, H.: Searching for Interacting Features. In: Proceedings of 20th International Joint Conference on Artificial Intelligence, Hyderabad, India (2007)

Discovering Proximal Social Intelligence for Quality Decision Support

Yuan-Chu Hwang

Abstract. The concepts of proximity have been utilized for exploring both psychological and geographical incentives for users within social networks to collaborate with others for mutual goals. The massive information does not facilitate quality decision support. In this chapter, we focus on discovering the proximal social intelligence for quality decision support. The utilization of investigating both the context and the content of the application domain from social network relationships would highly improve the information quality for better decisions.

Discovering proximal social intelligence from user's personal context they encountered enable the improvement of decision-making quality. We illustrate a case of leisure recommendatory e-service for bicycle exercise entertainment in Taiwan. We introduce the proximity e-service as well as its theoretical support. The most recent personalized experience according to its context provides remarkable perceptual data from unique information sources. Moreover, the social network relationships extend the power of the unique perceptual information to converge as the collective social network intelligence.

1 Introduction

The debate of "Content is King" least for a long time. But the content in leisure entertainment industry is still weak. The leisure entertainment content is usually monopolized by business owners, available information are bundled with marketing strategy that lay particular stress on specific commercial firm. Sometimes the quality of obtainable leisure entertainment information is insufficient for user to make equitable decisions. In order to improve the decision quality, appropriate reference materials should be provided for user to make fair judgments.

In order to improve the quality of content, possible solution including broaden the reference information from various feasible sources; retrieve from both homogeneous and heterogeneous information sources; gather information from user's social network relationships instead of the traditional sources. By focus on user's

Yuan-Chu Hwang
Department of Information Management, National United University, Taiwan
No. 1, Lien Da, Kung-Ching Li, Miao-Li 36003, Taiwan
e-mail: ychwang@nuu.edu.tw

I.-H. Ting et al. (Eds.): Mining and Analyzing Social Networks, SCI 288, pp. 125–137.
springerlink.com

essential needs such as perceptual feeling and their context, the analysis extends to the content information as well as the context of users. Moreover, we would like to explore the collaborative social network intelligence for quality decision results.

Recently, the number of user generated contents (UGC) in social media has been increasing rapidly. Ordinary people have now become producers of digital contents as well as consumers. They are capable of publishing their own contents and opinions on the social media such as FaceBook and Youtube. According to the definition of Wikipedia, collaboration is a recursive process where two or more people or organizations work together toward an intersection of common goals [20]. The information obtained from different social media sources may provide critical and essential information for user to make decisions. Since collaboration does not require leadership and can sometimes bring better results through decentralization and egalitarianism [18]. Users may strengthen their ability from various information sources of the social networks, including both heterogeneous and homogeneous social network relationships. Therefore, the social networks could reserve huge valuable information sources, and worthy for advanced utilization.

2 Proximal Social Network Intelligence

While social networks may contain abundant information for further utilization, but altruism between unfamiliar strangers is rarely seen. For the sake of increasing collaboration opportunities, there should be some incentives or stimulus that will increase the possibility of altruistic behaviors. There exist some psychological barriers that influence user's mind for contributing their ability for group's benefits. However, those barriers could be overcome by certain mental encouragements.

Classic social science studies long ago demonstrated that proximity frequently increases the rate of individuals communicating and affiliating in organizations and communities [1, 5]. Proximity also develops strong norms of solidarity and cooperation. Sociologists and anthropologists have long recognized that people can feel close to distant others and develop common identities with distant others who they rarely or never meet [2, 10]. Besides geographical distance, proximity places increased emphasis on individual homophily personal characteristics. The principle of homophily provides the basis for numerous social interaction processes. The basic idea is simple: "people like to associate with similar others" [3, 11, 15]. In this chapter, we utilize the concept of proximity to explore the social network intelligence and stress the collective efforts of participants in the dynamic environment. Homophylic user groups are more likely to combine the strength of different individuals to achieve specific objectives.

On the basis of the proximity concept, interpersonal social relationships could become vital information source with plentiful social energy for altruistic behaviors. The interpersonal social relationships can be defined by tie strength as weak or strong ties based on the following combinations: time, emotional intensity, intimacy, and the reciprocal services which characterize the tie. [7] According to Marsden and Campbell, tie strength depends on the quantity, quality, and frequency of knowledge exchange between actors, and can vary from weak to strong.

Stronger ties are characterized by increased communication frequency and deeper, more intimate connections. However, weak ties tend to link individuals to other social worlds, providing new sources of information and other resources [8]. Their very weakness means that they tend to connect people who are more socially dissimilar than those connected via strong ties. Weak ties contribute to social solidarity; community cohesion increases with the number of local bridges in a community [7]. According to Friedkin [6] the mix of weak and strong ties increases the probability of information exchange, and tends to comprise social network intelligence for collaborations.

2.1 Exploring the Social Context

As mentioned in previous section, this chapter focuses on discovering proximal social intelligence from leisure service participants to obtain useful information so as to improve decision quality. Since making decision is related to personal perception and the circumstance people belonging to. Previous research found that social context and the decision strategy affected decision acceptance, understanding, decision time, and affective reactions to the group [19]. Consequently, in this chapter, both content data and context information from user's social network relationships will be utilized as diversely information sources, including their heterogeneous and homogeneous social network structures.

Social context of an individual is the culture that he or she was educated and/or lives in, and the people and institutions with whom the person interacts [21]. Social context reflects how the people around something use and interpret it. The social context influences how something is viewed. Personal experience could be various from different social context they encountered. Even when participate in the same event, the social context may influence people's perception and result in different experiences. For example, when watching a movie at the theater with friends, the feeling would be quite different than watching a movie provided from our boss for propaganda and education. Seeing a movie with friends look more joyful than the other that boss may require us to do more analysis and tasks. Depending on the social context we encountered, the gained experience will be quite different.

However, from the proximity perspective, people from proximal social network are more likely to form a cooperative behaviors since they may have similar believing and values. The social context of leisure entertainment participants is likely to feel solidarity of its members, who are more likely to stay together, trust and help each other. Members of the same social context will often think in similar styles and patterns even when their conclusions differ [21]. In this chapter, leisure entertainment participants are encouraged to provide their personal experience for reference. By gathering updated and proximal leisure information, the provided service could benefit from those timely, relevance, and personal experience for further utilization.

Owing to the dynamicity and complexity present in our world, it is unrealistic to expect humans to be able to reason and act effectively to devote themselves for a collaborative environment. According to Maier, 1970, the results generated from user groups may induce greater acceptance of decisions [13]. The proximal social context enables relevant information exchange that may also provide some clues that draw user's attraction. The assertiveness and achievement of contributors would also become the essential incentives for user to collaborate with proximal others.

The remaining sections are shown as follows. In section 3, we explore both context information and content data from leisure entertainment participants. The TF/IDF and CTD (Category Term Descriptor) methods for leisure information are introduced and applied for recommendatory service. We introduce a leisure entertainment recommendatory e-service that is designed based on the proximal social intelligence in section 4. The evaluation of the recommendatory methods and managerial implications are shown in section 4, too. Finally, a conclusion and future directions of our work are provided in Section 5.

3 Exploring the Proximal Social Intelligence

Social network intelligence reserves rich personal information according to user's social context. If the reserved information is utilized properly, users can obtain important information from user peers within the same social context for quality decision. The appropriate utilization of this collective intelligence could leads to extensive knowledge enhancement for its domain. Shops and government can utilize those information for improving their provided product and service. Customers could also benefit from other customer's opinions, thus form a collaborative and healthy context environment. In the collaborative leisure recommendatory service, users can devote their up-to-the-minute personal experience as the input of the service. The provided personal experience are deposited in text format and stored as a tag. By gathering personal feedbacks acquired from the proximal social context for progressive mining technique, the leisure recommendatory service will obtain tremendous quality information for user to improve the overall decision quality.

In this chapter, we provide a leisure recommendatory e-service that allows the users to provide representative description of their perceptual experience regarding the leisure related events they encountered. Next, we use these user's perceptional descriptions as the hints to introduce the target event. The leisure entertainment participants can review certain initial concepts from others with similar social context. The provided tags are presented according to different methods to deliver an overview for specific target.

Two collaborative text mining techniques are applied in this leisure recommendatory service. The TF-IDF (Term Frequency-Inverse Document Frequency) and CTD (Category Term Descriptor) are utilized for extract useful personal feedback information for user to shape their knowledge and improve the decision quality. The two methods are elaborated as follows.

3.1 The TF-IDF Method

Term Frequency Inversed Document Frequency, or abbreviate as TF-IDF is one of the most popular term weighting schemes in information retrieval. The concept of Inverse Document Frequency (IDF) was proposed by Spark Jones, K in 1972 for explaining the statistical significance of the keywords [17]. Term Frequency (TF) was proposed by Salton & McGill in 1983 aims for data indexing with IDF, which integrate TF and IDF and become the a weighting algorithm for the keywords. The reason to use this algorithm is that the keywords used in each document vary from document to document [16] and therefore by combining TF and IDF, it is now possible to derive the relative weight of a keyword in all documents.

TF-IDF is mainly used in finding the relative weight of a keyword in a document. TF means the frequency of appearance of the keyword and IDF is used to find the relative importance of the keyword.

$$IDF_i = \log \frac{|D|}{|\{d_j : t_i \in d_j\}|}$$

$|D|$ is the total number of documents

$|\{d_j : t_i \in d_j\}|$ is the number of documents that contains the keyword $\boldsymbol{i}$.

$TF_{i,j} = \frac{n_{i,j}}{\sum_k n_{k,j}}$ denote the frequency count of the appearance of a keyword in a document divided by sum of all keywords' appearance frequency.

TF shows the relative importance of the keywords in a given document. IDF shows the importance of this keyword in the entire cohort. A keyword will be given higher IDF value if it is used only in small number of documents because it has more discriminative power.

For example, in the cultural event, if the word "Hakka"(a unique ethnic group of "Han" Chinese) is considered a keyword and it appears in a small number of documents, its IDF value would be high. However, the words like "food" and "good" appear in all documents and therefore have the IDF value close to zero. In TF, the more frequently a word is used, the higher the TF value in relation to the total number of keywords in a document. If the word "Hakka" is used in a document frequently, since it has high IDF and high TF, the word "Hakka" should be considered a very significant keyword for recommendation.

This method of utilizing the tags from heterogeneous information sources leads to research issues for tag classification and weighting. As described above, tags with high frequency count does not necessarily mean it is more important, therefore we will classify the tags using TF-IDF algorithm to provide accurate result for decision reference.

3.2 The CTD Method

The CTD (Category Term Descriptor) method was proposed by Bong & Narayanan in 2004. It is derived based on classic term weighting scheme, TF-IDF. The method explicitly chooses feature set for each category by only selecting set of terms from relevant category. Authors of CTD claim that incorporating only relevant feature can be highly effective and perform comparatively well with other measures, especially on collection with highly overlapped topics [9]. Since the leisure entertainment event could be unfolded in several categories for comprehensible description. We utilize CTD method as alternative comparative method for providing reference information for recommendation. Since the decision quality is subjective to user's perception and their social context, the original performance measuring matrix is replaced by cognition parameters in this chapter.

The proposed CTD method is extend from TF-IDF, where TF refers to term frequency in category ***c*** and ICF is interpreted as inverse category frequency. TF-ICF scheme shows no way of discriminating between terms that occur frequently in a small subset of documents and terms that are present in a large number of documents throughout a category. The formula of CTD was defined as follows.

$$CTD(t_k, c) = TF(t_k, c_i) \cdot IDF(t_k, c_i) \cdot ICF(t_k)$$

where

$$ICF(t_k) = \log\left(\frac{|C|}{CF(t_k)}\right)$$

$$IDF(t_k, c_i) = \log\left(\frac{|D(c_i)|}{DF(t_k, c_i)}\right)$$

$D(c_i)$ denote the number of document in category c_i

C denote number of category in the collection.

$CF(t_k)$ denote the category frequency for term t_k

$DF(t_k, c_i)$ denote the document frequency for term t_k in category c_i

The CTD method also utilizing the tags from heterogeneous information sources for tag classification and weighting. For the purpose of taking the classification issue into consideration, CTD method is also used in this chapter for contribute proximal social intelligence for leisure recommendatory service.

4 i-Bike Leisure Recommendatory Service

Based on the concept of proximity from social network relationships, we propose a collaborative leisure entertainment recommendatory service, called "i-Bike". The i-Bike service explores those proximal social intelligence from both context and content and enable quality decision making. The i-Bike service illustrates and exchanges user's personal experience for bicycle exercise entertainment in Taiwan. A schematic diagram of interactions is shown as Figure 1. The process can be unfolded into two parts, one is the experience contribution process, and the other is knowledge acquisition process.

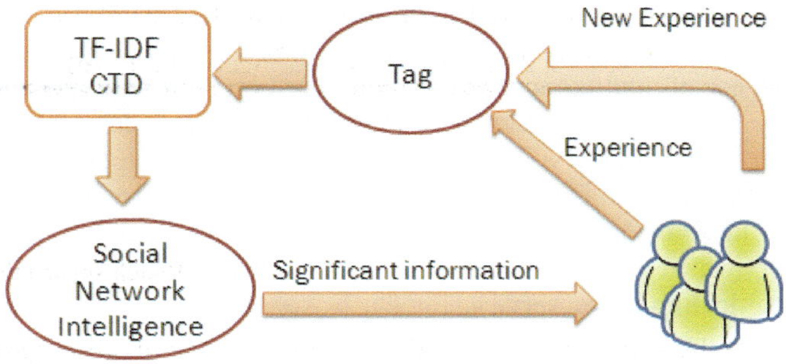

Fig. 1 Schematic interaction process of i-Bike service

The contribution process is a spontaneous action that users are encouraged to share their experience in text format after their tour events. For each bicycle tour spot, the service platform allows users to contribute their personal experience and preserve in database for utilization. According to the previous mentioned methods, the most recent and important personalized experience from the context that user belong to is generated automatically and ready for operation. The knowledge acquisition process is very simple. Users can access the leisure entertainment recommendatory service and retrieve remarkable perceptual data from unique information sources. The proximal social network relationships converges every unique perceptual experience as the collective social network intelligence for improving the decision making process. A system sketch is shown as Figure 2, the right parts that contain both TF-IDF and CTD results which represent social network intelligence for i-Bike leisure entertainment recommendatory service.

Fig. 2 Sketch of iBike Leisure Entertainment Recommendatory Service

4.1 Measuring the Decision Quality of i-Bike Service

The proximal social intelligence is generated based on collective wisdom, which is motivated by psychological incentives. However, the free riders may exist in the service environment as well. The altruistic behaviors still happened which is encouraged by the attraction of similar interests and the assertiveness and achievement from others. In order to measure the decision quality improvement of our proposed service, the measuring matrix must focus on psychological parameters and perceived feeling of the service.

The decision quality is subjective measures. It evaluates how user satisfied with their decisions. According to Lilien et.al, subjective measures can provide additional valuable insights into decision effectiveness [12]. It is particularly useful for assessing consumer evaluations of the decision process and their feelings of the decision.

We utilize some perceptual measure parameters for evaluating the decision quality after utilizing i-Bike recommendatory service. The measuring parameters in this chapter is are unfolded into five dimensions, including perceived usefulness, perceived easy to use, information quality satisfaction, decision result satisfaction, and willingness to contribute. We utilize these dimensions for evaluating both the TF-IDF and CTD methods comparing to traditional TF method. The questionnaire for measuring the impact of i-Bike recommendatory service system is designed according to technology acceptance model (TAM) [4] We also evaluate the difference between TF-IDF and CTD methods and analysis the free-rider issue in our research. The questionnaire contains sixteen questions and uses a 7-points Likert Scale measurement design. The reliability analysis of the questionnaire indicates the Cronbach's α is 0.835 and the split-half reliability is 0.822. The Cronbach'sα is higher than 0.70 which indicates that the measure is reliable [14].

A randomized control trial (RCT) was applied to this study. The subjects were recruited from university students and they were randomly assigned into each group. Two 32 users teams in experimental group 1 and 2, and another 32 users in control group completed the study. Users in the experimental group 1 received the leisure information generated from TF-IDF method, users in the experimental group 2 received the information from CTD method, and users in the control group received the information generated from traditional TF method. Users will receive leisure entertainment information from different recommendation mechanisms. The provision of leisure entertainment information includes the desired travel route of bicycle of Maio-Li County in Taiwan. Each route will presented in the geographic map and indicate most recommended spots in each route. The recommend information is presented into tags that are generated from other users. Each spot will be presented with pictures in users screen and the GPS information of the spot will also provided to users.

In each group, subjects received recommendatory service information and presented in a form of tags. The recommend mechanisms of each experiment are different, but the information layout of the recommendatory service is in the same way. In this experiment, the provided information presented in each spot contains 10 most useful description tags computed in different recommend methods. Users can evaluate the provided recommend service and the provided information first. Users are also allowed to provide description information of each spot using information tags. These tags will be included in the database and for future utilization. In this experiment, the tag information in database is previously generated from students who had visited the spots of each bicycle route. During the experiment process, the information input function is temporary disconnected to main database so as to make sure users will receive recommendations from the same database. Nevertheless, their contributions are manually added into database for further analysis. Again, the only difference in each experiment is the recommend mechanism. We will examine and analysis differences between the three recommend methods according to the five evaluate dimensions. Following are the experiment results of each paired group comparison.

4.2 Experiment Result

We use the independent t-test to examine the difference between each paired group. The decision quality comparisons include three pairs: (1) TF-IDF method vs. TF method, (2) CTD method vs. CTD method, and (3) TF-IDF method vs. CTD method.

When comparing the TF-IDF method with traditional TF method, there was a significant difference between the two groups. This significant difference between two groups was confirmed by independent t-test ($p<0.05$) in each dimensions. The TF-IDF method provides high quality information for users than traditional recommend method. Users feel the recommend information using TF-IDF could provide high quality information to improve their decision. Unsurprisingly, using term frequency (TF) could only bring the popular issues for user. However, the provided information may be too conceptual that did not addressing user's need for detailed entertainment information.

The comparing outcome of CTD method with TF method is similar to previous experiment. The independent t-test ($p<0.05$) indicate the CTD recommendation provide high quality information than the TF method in all five dimensions. We can see the information using CTD is more addressing the category of the leisure information than TF method.

However, when comparing the TF-IDF method with CTD method, there is no significant difference between the two recommending mechanism in all dimensions. We will explore the behavioral analysis in this issue in the following subsection.

In summary, when compare with using pure term frequency for recommendation, both the TF-IDF and CTD methods are comparatively successful than traditional TF method. The proximal social intelligence brings latest information for each leisure e-service users. Moreover, the provided information contains personal opinions and suggestions that are not available from traditional information sources. The i-Bike service provide a platform for users to collaborate with others to filer and extract useful information that is focusing on user's urgent needs. The satisfaction of information quality in each recommendatory mechanism is improved than using TF method.

4.3 User Behavior and Free Rider Issue

Previous experiment results indicate the two methods (TF-IDF and CTD) could provide satisfactory outcomes than traditional TF method. However, the difference between TF-IDF and CTD is not significant. In order to analysis the reason, we go further to analysis the user behaviors when using i-Bike service.

According to the contribution rate in user's behavior log file, we found the user behaviors could be unfolded into three types: active users, passive users, and free riders. Active users are who will contribute information and provide feedback information energetically; there are about 13% of active users in the service system. Passive users will still contribute information and feedbacks; however, the quantity of information is comparatively less than active users. In this study, there are 27% of passive users in the service system. We found there is a significant free rider problem in the service system. There are 40% users will contribute to the system positively, but other user's behavior are shown as free riders. These free riders do consider the provided proximity social intelligence for improving the decision quality. However, they did not contribute or provide feedback information to the service system.

Since i-Bike service is a Web 2.0 service that addressing on the concept of user generated content, the less contribution may imply user's attitude to collaborative behavior. The proximal social intelligence do requires the contribution of all possible information sources to enable the collative wisdom. However, the less feed back contribution and less information sources could diminish the power and influence the impact of social intelligence. In order to understand the difference of the recommending mechanism, we evaluate the difference by exclude the free riders to determine the situations on those positive users.

After separate the free riders, analysis result indicate the TF-IDF method has better satisfaction than CTD method. There are difference between two groups was confirmed by independent t-test ($p<0.1$) in perceived usefulness dimension and information quality satisfaction dimension. We can say that users who positively contribute to the service system could experience varied difference of the two recommending mechanisms.

Although we found there exist some difference between TF-IDF and CTD recommendatory methods, but the variation is little. The original CTD method is proposed to improve the TF-IDF method, but the effect is not clear in our case. It could be influenced by the classification of the bicycle leisure entertainment that reduces the effect in this particular field domain.

4.4 Managerial Implications

The goal of this chapter is to explore the proximal social intelligence for collaborative recommendation. The design applied the web 2.0 service design that enables users participate in the collaboration process and empower the experience co-creation. Different recommendatory mechanisms were used for generating suitable tags to enrich information quality to improve the decision quality. According to the experiment result, both TF-IDF and CTD methods could provide high quality information for decision making than traditional way. However, we found that both effective recommendatory mechanism and the suitable atmosphere, which encourage users to contribute information actively, are critical for a successful recommendatory service.

The reason for users to contribute themselves to proximal social intelligence dependent on what user perceived from the provided e-service. The usefulness, easy to use, and high quality information are essential factors that encourage users participate in the service continuously. Appropriate recommendatory mechanism could improve the information quality, but to facilitate the participation rely on suitable design that focus on user's perception. Future research design should strengthen the incentives for user to share information actively. In order to provide some incentives or stimulus that increasing collaboration opportunities and facilitating the altruistic behaviors, some possible improvements are described below.

First, a suitable ranking system or point system should be established. Providing the differentiation of information provider could honor the active information providers and create incentives for others continuously contribute themselves to the service system. Second, some ad-hoc user group could be established according to the user's need and their interests. By gathering more users participate in the service, more unique information could be obtained and enrich the recommendatory capability. When users establishing some emotional connections, they will be lock-in the system and also urge other users to join the society. This will bring more valuable heterogeneous information sources and personal experiences into the service system to improve the information quality as well as the decision quality.

5 Conclusion and Future Directions

In web 2.0 era, collaboration with ubiquitous social intelligence becomes an important trend. The social networks contain abundant information for further utilization, but altruism between unfamiliar strangers is need urgently. This chapter focuses on user's basic needs such as perceptual feeling and their context, the analysis extends to the content information as well as the context of users. By highlight the proximity of each participant, we extend the user generated contents (UGC) in social media for better utilization. The concepts of proximity have been utilized for exploring both psychological and geographical incentives for users within social networks to collaborate with others for mutual goals.

Personal experience could be various from different social context they encountered. We utilize these unique information sources to improve the recommendation quality for leisure entertainment industry. Both TF-IDF and CTD methods are applied for extracting the knowledge from social network intelligence.

The collective social network intelligence benefit from user's proximity recognition provides strong incentives for user to contribute themselves for mutual advantages. The social network on cyberspace also provide significant information exchange rate. Utilizing the proximal social network intelligence on Internet environments, leisure entertainment recommendatory service may transmit information various ways. Through the external channel that permits effective information spread and diffusion. Therefore, the participants with similar interest could be encouraged to share and contribute their experiences with whom they encountered. Through the internal method, mining the proximal social network intelligences helps individuals to gather and obtain useful information via social network structures. Proximal participants would propagate information voluntarily via their own social networks voluntarily. Information diffusion for proximity e-services is more efficient and supply rich reference information for improving the decision quality.

For the direction of future research, some extend evaluation could be further examined: such as the social utility of mining the proximal social intelligence, the participation rate of users and the contribution quality analysis through various social network structures.

References

[1] Allen, T.J.: Managing the Flow of Technology. MIT Press, Cambridge (1977)

[2] Anderson, B.: Imagined Communities. Verso, London (1983)

[3] Aristotle: The Nicomachean Ethics. H. Rackham, translator. Harvard University Press, Cambridge (1934)

[4] Davis, F.D.: Perceived usefulness, perceived ease of use, and user acceptance of information technology. MIS Quarterly 13(3), 319–340 (1989)

[5] Festinger, L., Schachter, S., Back, S.: Social Pressures in Informal Groups: A Study of Human Factors in Housing. Stanford University Press, Palo Alto (1950)

[6] Friedkin, N.E.: Information flow through strong and weak ties in intraorganizational social networks. Social Networks 3, 273–285 (1982)

[7] Granovetter, M.: The Strength of Weak Ties. American Journal of Sociology 78, 1360–1380 (1973)

[8] Granovetter, M.: The strength of weak ties: A network theory revisited. In: Marsden, P.V., Lin, N. (eds.) Social Structure and Network Analysis, pp. 105–130. Sage, Beverly Hills (1982)

[9] How, B., Narayanan, K.: An Empirical Study of Feature Selection for Text Categorization based on Term Weightage. In: Proceedings of the IEEE/WIC/ACM International Conference on Web Intelligence 2004, pp. 599–602 (2004)

[10] Habermas, J.: The structural transformation of the public sphere. MIT Press, Cambridge (1991)

[11] Lazarsfeld, P., Merton, R.K.: Friendship as a social Process: A Substantive and Methodological Analysis. In: Berger, M., Abel, T., Page, C.H. (eds.) Freedom and Control in Modern Society, pp. 18–66. Van Nostrand, New York (1954)

[12] Lilien, G.L., Rangaswamy, A., Van Bruggen, G.H., Starke, K.: DSS Effectiveness in Marketing Resource Allocation Decisions: Reality vs. Perception. Information Systems Research 15, 216–235 (2004)

[13] Maier, N.R.F.: Problem Solving and Creativity; in Individuals and Groups. Brooks/Cole Publishing Company, Belmont (1970)

[14] Nunnally, J.C.: Psychometric Theory, 2nd edn. McGraw-Hill, New York (1978)

[15] Plato, Laws. Plato in twelve volumes. In: Bury translator, vol. 11, p. 837. Harvard U. Press, Cambridge (1968)

[16] Salton, G., McGill, M.J.: Introduction to modern information retrieval. McGraw-Hill, New York (1983)

[17] Sparck Jones, K.: A statistical interpretation of term specificity and its application in retrieval. Journal of Documentation (1972)

[18] Spence, Muneera, U.: Graphic Design: Collaborative Processes - Understanding Self and Others, 13 April 2006. Oregon State University, Corvallis (2006)

[19] Tjosvold, D., Field, R.H.G.: Effects of social context on consensus and majority vote decision making. Academy of Management Journal 26, 500–506 (1983)

[20] Wikipedia, `http://en.wikipedia.org/wiki/Collaboration` (Accessed October 13, 2009)

[21] Wikipedia, `http://en.wikipedia.org/wiki/Social_environment` (Accessed October 13, 2009)

Discovering User Interests by Document Classification

Loc Nguyen

User interest is one of personal traits attracting researchers' attention in user modeling and user profiling. User interest competes with user knowledge to become the most important characteristics in user model. Adaptive systems need to know user interests so that provide adaptation to user. For example, adaptive learning systems tailor learning materials (lesson, example, exercise, test…) to user interests. I propose a new approach for discovering user interest based on document classification. The basic idea is to consider user interests as classes of documents. The process of classifying documents is also the process of discovering user interests. There are two new points of view:

- The series of user access in his/her history are modeled as documents. So user is referred indirectly to as "document".
- User interests are classes such documents are belong to.

Our approach includes four following steps:

1. Documents in training corpus are represented according to **vector model**. Each element of vector is product of term frequency and inverse document frequency. However the inverse document frequency can be removed from each element for convenience.
2. **Classifying training corpus** by applying decision tree or support vector machine or neural network. Classification rules (weight vectors W^*) are drawn from decision tree (support vector machine). They are used as classifiers.
3. Mining user's access history to **find maximum frequent itemsets**. Each itemset is considered a interesting document and its member items are considered as terms. Such interesting documents are modeled as vectors.
4. Applying classifiers (see step 3) into these interesting documents in order to choose which classes are most suitable to these interesting documents. **Such classes are user interests**.

This approach bases on document classification but it also relates to information retrieval in the manner of representing documents. Hence section 1 discusses about vector model for representing documents. Support vector machine, decision tree and neural network on document classification are mentioned in section 2, 3, 4.

Loc Nguyen
University of Science, Ho Chi Minh city, Vietnam
e-mail: ng_phloc@yahoo.com

I.-H. Ting et al. (Eds.): Mining and Analyzing Social Networks, SCI 288, pp. 139–159.
springerlink.com

Main technique to discover user interest is described in section 5. Section 6 is the evaluation.

1 Vector Model for Representing Documents

Suppose our corpus **D** is the composition of documents $D_i \in \mathbf{D} = \{D_1, D_2, \ldots, D_m\}$. Every document D_i contains a set of key words so-called *terms*. The number of times a term occurs in a document is called *term frequency*. Given the document D_i and term t_j, the term frequency tf_{ij} measuring the importance of term t_j within document D_i is defined as below:

$$tf_{ij} = \frac{n_{ij}}{\sum_k n_{ik}}$$

Where n_{ij} is the number of occurrences of term t_j in document D_i, and the denominator is the sum of number of occurrences of all terms in document D_i.

Suppose we need to search documents which are most relevant to the query having term t_j. The simple way is to choose documents which have highest term frequency tf_{ij}. However in situation that t_j is not a good term to distinguish between relevant and non-relevant documents and other terms occurring rarely are better ones to distinguish between relevant and non-relevant documents. This will tend to incorrectly emphasize documents containing term t_j more, without giving enough weight to other meaningful terms. So the *inverse document frequency* is a measure of general importance of the term. It is used to decrease the weight of terms occurring frequently and increase the weight of terms occurring rarely. The inverse document frequency of term t_j is the ratio of the size of corpus to the number of documents that t_j occurs.

$$idf_j = \log \frac{|corpus|}{|\{D: t_j \in D\}|}$$

Where $|corpus|$ is the total number of documents in corpus and $|\{D: t_j \in D\}|$ is the number of documents containing term t_j. We use *log* function to normalize idf_j so that it is less than or equal *1*.

The weight of term t_j in document D_i is defined as product of tf_{ij} and idf_i

$$w_{ij} = tf_{ij} * idf_i$$

This weight measure the importance of a term in a document over the corpus. It increases proportionally to the number of times a term occurs in the document but is offset by the frequency of this term in the corpus. In general this weight balances the importance of two measures: term frequency and inverse document frequency.

Suppose there are n terms $\{t_i, t_2, \ldots, t_n\}$, each document D_i is modeled as the vector which is composed of weights of such terms.

$$D_i = (w_{i1}, w_{i2}, w_{i3}, \ldots, w_{in})$$

Hence the corpus becomes a matrix m x n, which have m rows and n cols with respect to m document and n terms. D_i is called document vector.

The essence of document classification is to use supervised learning algorithms in order to classify corpus into groups of documents; each group is labeled. In this chapter we apply two methods namely support vector machine, decision tree and neural network for document classification.

2 Document Classification Based on Support Vector Machine

2.1 Support Vector Machine

Support vector machine (SVM) [Cristianini, Shawe-Taylor 2000] is a supervised learning algorithm for classification and regression. Given a set of n-dimensional vectors in vector space, SVM finds the separating hyper-plane that splits vector space into sub-set of vector; each separated sub-set (so-called data set) is assigned one class. There is the constraint for this separating hyper-plane: "it must maximize the margin between two sub-sets".

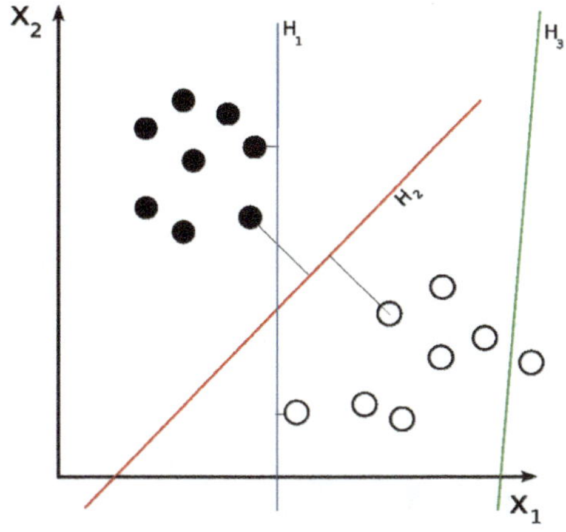

Fig. 1 Separating hyper-planes

Suppose we have some n-dimensional vectors; each of them belongs to one of two classes. We can find many n-1 dimensional hyper-planes that classify such vectors but there is only one hyper-plane that maximizes the margin between two classes. In other words, the nearest between a point in one side of this hyper-plane and other side of this hyper-plane is maximized. Such hyper-plane is called maximum-margin hyper-plane and it is considered as maximum-margin classifier.

Let $\{X_1, X_2,..., X_n\}$ be the training set of vectors and let $y_i = \{1, -1\}$be the class label of vector X_i. It is necessary to determine the maximum-margin hyper-plane that separates vectors belonging to y_i=1 from vectors belonging to y_i = -1. This hyper-plane is written as the set of point satisfying:

$$W^T \otimes X_i + b = 0 \quad (1)$$

Where $\otimes$ denotes the scalar product and W is a weight vector perpendicular to hyper-plane and b is the bias. W is also called perpendicular vector or normal vector. It is used to specify hyper-plane.

The value $\frac{b}{|W|}$ is the offset of the hyper-plane from the origin along the weight vector W.

To calculate the margin, two parallel hyper-planes are constructed, one on each side of the maximum-margin hyper-plane. Such two parallel hyper-planes are represented by two following equations:

$$W^T \otimes X_i + b = 1$$
$$W^T \otimes X_i + b = -1$$

To prevent vectors falling into the margin, all vectors belonging to two class $y_i=1$, $y_i=-1$ have two following constraints respectively:

$$W^T \otimes X_i + b \geq 1 \quad \text{(for } X_i \text{ of class } y_i=1\text{)}$$
$$W^T \otimes X_i + b \leq -1 \quad \text{(for } X_i \text{ of class } y_i=-1\text{)}$$

These constraints can be re-written as:

$$Y_i\,(W^T \otimes X_i + b) \geq 1 \quad (2)$$

For any new vector X, the rule for classifying it is computed as below:

$$f(X_i) = \text{sign}(W^T \otimes X_i + b) \in \{\leq -1, \geq 1\} \quad (3)$$

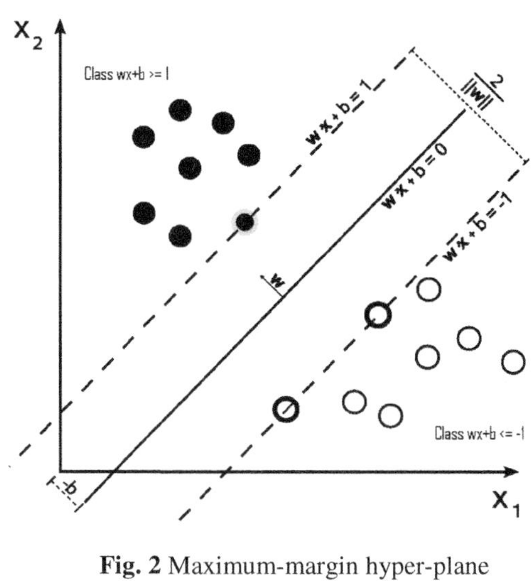

Fig. 2 Maximum-margin hyper-plane

Because maximum-margin hyper-plane is defined by weight vector W, it is easy to recognize that the essence of constructing maximum-margin hyper-plane is to solve the following constrained optimization problem:

$$\underset{W,b}{Minimize}\ \frac{1}{2}|W|^2 \text{ subject to } y_i\,(W^T \otimes X_i \ + \ b) \geq 1,\ \forall i$$

This constraints can be re-written as below:

$$\underset{W,b}{Minimize}\ \frac{1}{2}|W|^2 \text{ subject to } 1 - y_i\,(W^T \otimes X_i \ + \ b) \leq 0,\ \forall i \tag{4}$$

The reason of minimize $\frac{1}{2}|W|^2$ is that distance between two parallel hyper-planes is $\frac{2}{|W|}$ and we need to maximize such distance in order to maximize the margin for maximum-margin hyper-plane. Then maximizing $\frac{2}{|W|}$ is to minimize $\frac{1}{2}|W|$. Because $|W|$ is the norm of W being complex to compute, we substitute $\frac{1}{2}|W|$ with $\frac{1}{2}|W|^2$. This is the *constrained optimization problem* or *quadratic programming* (QP) *optimization* problem. Note that $|W|^2$ can be computed by scalar product of it and itself:

$$|W|^2 = W^T \otimes W$$

The way to solve QP problem in formula III.3.2 is through its Lagrange dual. Suppose we want to minimize $f(x)$ subject to $g(x)=0$, it exists a solution x_0 to set of equations:

$$\begin{cases} \frac{\partial}{\partial x}(f(x)+\alpha g(x))_{x=x0} = 0 \\ g(x)=0 \end{cases}$$

where α is Lagrange multiplier.

For multi constraints $g_i(x) = 0$ ($i=\overline{1,n}$), there is a Lagrange multiplier for each constraint

$$\begin{cases} \frac{\partial}{\partial x}(f(x)+\sum_{i=1}^{n}\alpha_i g_i(x) \Big|_{x=x0} = 0 \\ g_i(x)=0 \end{cases}$$

In case of $g_i(x) \leq 0$, there is no change. If x_0 is a solution to the constrained optimized problem:

$$\underset{x}{Minimize}\ f(x)\ subject\ to\ g_i(x) \leq 0\ \forall i$$

Then there must exist $\alpha_i \geq 0\ \forall i$ so that x_0 satisfies following equations:

$$\begin{cases} \frac{\partial}{\partial x}(f(x)+\sum_{i=1}^{n}\alpha_i g_i(x) \Big|_{x=x0} = 0 \\ g_i(x) \leq 0 \end{cases} \tag{5}$$

The function $f(x)+\sum_i \alpha_i g_i(x)$ is called Lagrangian denoted L. Let $f(X_i)$ be $\frac{1}{2}|W|^2$ and let $g_i(X)$ be $1-y_i(W^T \otimes X_i + b)$, the constraint in formula III.3.4 becomes:

$$\underset{\lambda}{Maximize}(\underset{W,b}{Minimize} L(W,b,\lambda)) \tag{6}$$

where

$$L(W,b,\lambda) = \frac{1}{2}|W|^2 + \sum_{i=1}^{n}\lambda_i(1-y_i(W^T \otimes X_i + b)) \tag{7}$$

The Lagrangian L is minimized with respect to the primal variables W and b and maximized with respect to the dual variables λ. Setting the gradient of L w.r.t W and b to zero, we have

$$\begin{cases} \frac{\partial L(W,b,\lambda)}{\partial W} = 0 \\ \frac{\partial L(W,b,\lambda)}{\partial \lambda} = 0 \end{cases} \Leftrightarrow \begin{cases} W - \sum_{i=1}^{n}\lambda_i y_i X_i = 0 \\ \sum_{i=1}^{n}\lambda_i y_i = 0 \end{cases} \tag{8}$$

Substituting (III.3.8) into (III.3.7) we have:

$$L(\lambda) = \sum_{i=1}^{n}\lambda_i - \frac{1}{2}\sum_{i=1}^{n}\sum_{j=1}^{1}\lambda_i\lambda_j y_i y_j X_i^T X_j \tag{9}$$

Where $\lambda = (\lambda_1, \lambda_2, \ldots, \lambda_n)^T$.

If (9) is re-written in matrix notation then the constrained optimization problem in (4), (6) leads to dual problem:

$$Maximize\ L(\lambda) = \lambda^I I - \frac{1}{2}\lambda^T S\lambda\ subject\ to\ \lambda \geq 0\ and\ \lambda^t y = 0 \tag{10}$$

Where S is a symmetric $n_x n$ matrix with elements $s_{ij}=y_i y_j X_i^T X_j$.

Suppose λ^* is a solution of (III.3.1) with condition (III.3.8). In order words, $L(\lambda^*)$ is maximized; so the weight vector representing maximum-margin hyper-plane is recovered by λ^* and X_i:

$$W^* = \sum_{i=1}^{n} \lambda_i y_i X_i \tag{11}$$

So the bias b is computed as below:

$$b^* = y_i - W^{*T} \otimes X_i \tag{12}$$

The rule for classification in (3) becomes:

$$f(X_i) = R = \text{sign}(W^{*T} \otimes X_i + b^*) \tag{13}$$

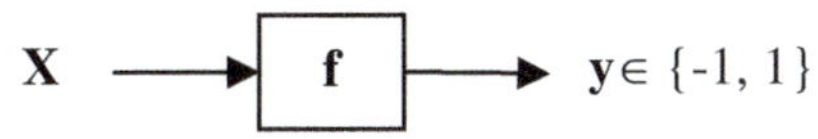

Fig. 3 Classification function

It means that whenever we need to determine to which class a new vector X_i belongs, it is only to substitute X_i into $W^{*T} \otimes X_i + b^*$ and check the value of this expression. If the value is less than or equal *-1* (≤ -1) then X_i belongs to class $y_i = -1$. Otherwise, if the value is greater than or equal *1* (≥ 1) then X_i belongs to class $y_i = 1$. Hence the function ($W^{*T} \otimes X_i + b^*$) is called classification function or classification rule.

The Lagrange multipliers are non-zero when $W^T \otimes X_i + b$ is equal *1 or -1*, vectors X_i in this case are considered support vectors they are closest to the maximum-margin hyper-plane. These vectors lie on parallel hyper-planes. So this approach is called support vector machine.

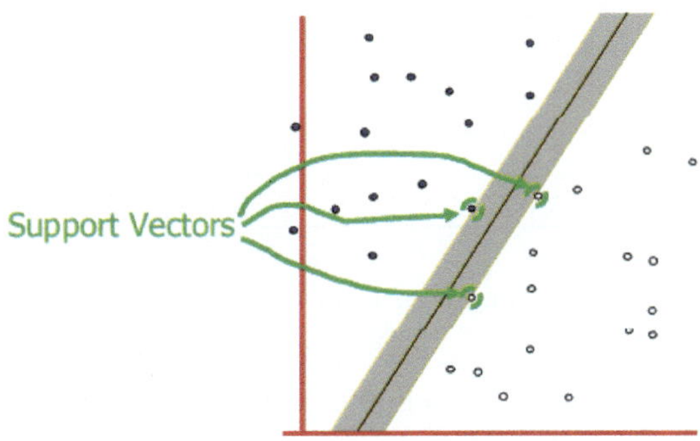

Fig. 4 Support vectors

However the way to find λ^* in (10) is quadratic programming (QP) problem. There are many approaches to sole this problem; for instance, the sequential minimal optimization is described as below:

- A QP with two variables is trivial to solve
- At each iteration, a pair of (λ_i, λ_j) is picked up and QP is solved with these variables; repeat until convergence

2.2 Applying Support Vector Machine into Document Classification

Give corpus $\mathbf{D} = \{D_1, D_2, D_3,..., D_m\}$ in which every document D_i is modeled by a *tf-idf* weight vector. Suppose there are n terms $\{t_1, t_2,..., t_n\}$, we have:

$$D_i = (d_{i1}, d_{i2},..., d_{in})$$

where d_{ij} is product of term frequency and inverse document frequency $d_{ij} = tf_{ij}*idf_j$.

If the index of document is ignored, document D is represented as below:

$$D = (d_1, d_2,..., d_n)$$

Given k classes $\{l_1, l_2,..., l_k\}$, there is demand of classifying documents into such classes. The technique of classification based on SVM is *two-class* classification in which the classes is +1, -1 for $y_i=+1, -1$ respectively. So we need to determine unique hyper-plane referred to as *two-class* classifier. It is possible to extend *two-class* classification to *k-class* classification by constructing k *two-class* classifier. In means that we must specify k couple of optimal weight vector W_i^* and bias b_i^*. Each couple (W_i^*, b_i^*) being a *two-class* classifier is the representation of class l_i. The process of finding (W_i^*, b_i^*) in training corpus **D** is described in the section of support vector machine.

Table 1 k couple (W_i^*, b_i^*) corresponds with k class $\{l_1, l_2,..., l_k\}$

Class	Weight vector	Bias	Classification rule
l_1	W_1^*	b_1^*	$R_1 = W_1^{*T} \otimes X + b_1^*$
l_2	W_2^*	b_2^*	$R_2 = W_2^{*T} \otimes X + b_2^*$
...	...	...	...
l_k	W_k^*	b_k^*	$R_k = W_k^{*T} \otimes X + b_k^*$

For example, classifying document $D= (d_1, d_2,..., d_n)$ is described as below:

1. For each classification rule $R_i = W_i^{*T} \otimes X + b_i^*$, substituting each D into such rule. It means that vector X in such rule is replaced by document D.
   $Expression_i = W_i^{*T} \otimes D + b_i^*$
2. Suppose there is a sub-set of rules $\{R_{h+1}, R_{h+2},..., R_{h+r}\}$ that the value of expression $Expression_i = W_i^{*T} \otimes D + b_i^*$ is greater than or equal 1. We can

conclude that document D is belongs to r classes $\{l_{h+1}, l_{h+2}, \ldots, l_{h+r}\}$ where $\{l_{h+1}, l_{h+2}, \ldots, l_{h+r}\} \subseteq \{l_1, l_2, \ldots, l_k\}$.

Table 2 Classifying document D

Classification Expression	**Value**
$W_1^{*T} \otimes D + b_1^*$	$\geq 1 : D \in l_1$ $\leq -1 : D \notin l_1$
$W_2^{*T} \otimes D + b_2^*$	$\geq 1 : D \in l_2$ $\leq -1 : D \notin l_2$
…	…
$W_k^{*T} \otimes D + b_k^*$	$\geq 1 : D \in l_k$ $\leq -1 : D \notin l_k$

3 Document Classification Based on Decision Tree

Given a set of classes **C** = {*computer science, math*}, a set of terms **T** = {*computer, programming language, algorithm, derivative*} and the corpus **D** = {*doc1.txt, doc2.txt, doc3.txt, doc4.txt, doc5.txt*}. The training data is showed in following table in which cell (*i, j*) indicates the number of times that term *j* (column *j*) occurs in document *i* (row *i*).

Table 3 Term frequencies of documents

	computer	*programming language*	*algorithm*	*derivative*	***class***
doc1.txt	5	3	1	1	computer
doc2.txt	5	5	40	5	math
doc3.txt	20	5	20	55	math
doc4.txt	20	55	5	20	computer
doc5.txt	15	15	4	0.3	math
doc6.txt	35	10	45	10	computer

Table 4 Normalized term frequencies

	computer	*programming language*	*algorithm*	*derivative*	**class**
doc1.txt	0.5	0.3	0.1	0.1	computer
doc2.txt	0.05	0.05	0.4	0.5	math
doc3.txt	0.2	0.05	0.2	0.55	math
doc4.txt	0.2	0.55	0.05	0.2	computer
doc5.txt	0.15	0.15	0.4	0.3	math
doc6.txt	0.35	0.1	0.45	0.1	computer

Because the expense of real number computation is so high, all term frequencies are changed from real number into nominal value:

1. $0 \leq frequency < 0.2$: *low*
2. $0.2 \leq frequency < 0.5$: *medium*
3. $0.5 \leq frequency$: *high*

Table 5 Nominal term frequencies

	computer	*programming language*	*algorithm*	*derivative*	**class**
doc1.txt	high	medium	low	low	computer
doc2.txt	low	low	medium	high	math
doc3.txt	medium	low	medium	high	math
doc4.txt	medium	high	low	medium	computer
doc5.txt	low	low	medium	medium	math
doc6.txt	medium	low	medium	low	computer

The basic idea of generating decision tree [Mitchell 1997] is to split the tree into two sub-trees at the most informative node. Such node is chosen by computing its entropy or information gain. Following figure shows the decision tree generated from our training data.

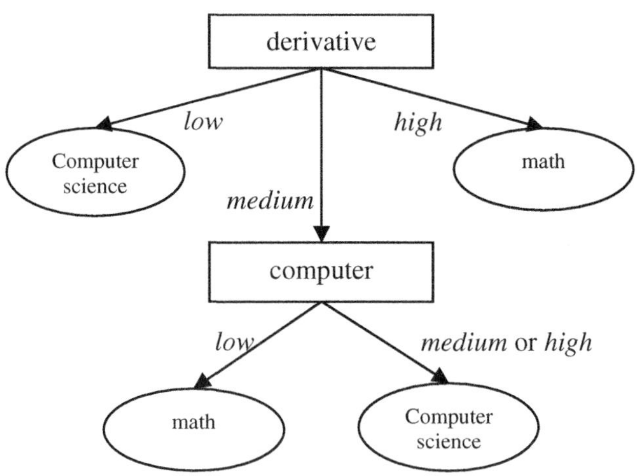

Fig. 5 Decision tree

We can extract classification rules from this decision tree:

Table 6 Classification rules deriving from decision tree induction

Rule 1	If frequency of term “derivative” is *low* then document belongs to class *computer science*
Rule 2	If frequency of term “derivative” is *medium* and frequency of term “computer” is *medium* or *high* then document belongs to class *computer science*
Rule 3	If frequency of term “derivative” is *medium* and frequency of term “computer” is *low* then document belongs to class *math.*
Rule 4	If frequency of term “derivative” is *high* then document belongs to class *math*

Suppose the numbers of times that terms *computer, programming language, algorithm* and *derivative* occur in document *D* are 5, 1, 1, and 3 respectively. We need to determine which class document *D* is belongs to. *D* is normalized as term frequency vector.

$D = (0.5, 0.1, 0.1, 0.3)$

Changing real number into nominal value, we have:

D = (*high, low, low, medium*)

According to rule 2 in above table, *D* is *computer science* document because in document vector *D*, frequency of term “derivative” is *medium* and frequency of term “computer” is *high*.

4 Document Classification Based on Neural Network

4.1 Artificial Neural Network

Artificial Neural Network (ANN) is the mathematical model based on biological neural network. It consists of a set of processing units which communicate together by sending signals to each other over a large number of weighted connections. Such processing units are also called neurons or cells or variables. Each unit is responsible for receiving input from neighbors or external sources and using this input to compute an output signal which is propagated to other units. However each unit also adjusts the weights of connections. There are three types of units:

- Input units receive data from outside the network. These units structure the input layer.
- Hidden units own input and output signals that remain within the neural network. These units structure the hidden layer. There can be one or more hidden layers.
- Output units send data out of the network. These units structure the output layer.

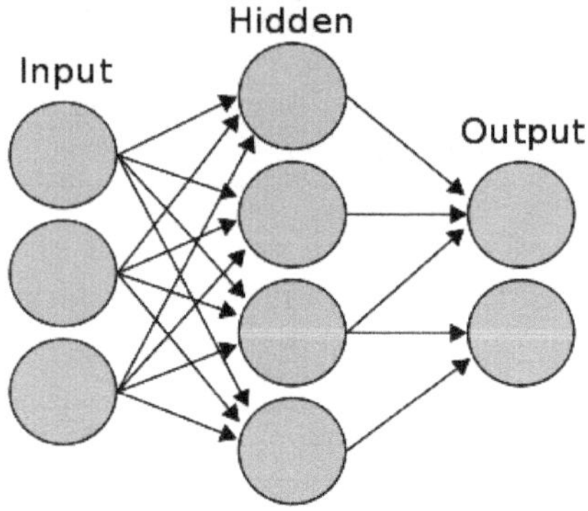

Fig. 6 Neural network

Each connection between unit i and unit j is defined by the weight w_{ij} determining the effect which the signal of unit i on unit j. Suppose input unit, hidden unit and output unit is denoted as x, h, y respectively. In the topology, unit y is the composition of other units h which in turn are the compositions of others units x. The composition of a unit is represented as a weighted sum which will be evaluated to determine the output of this unit. If such unit is the output unit, its output is the output of neural network. The process of computing the output of a unit includes two following steps:

- An adder so-called summing function sums up all the inputs multiplied by their respective weights. It is essential to compute the weighted sum. This activity is referred to as linear combination.
- An activation function controls the amplitude of the output of the neuron. This activity aims to determine the output.

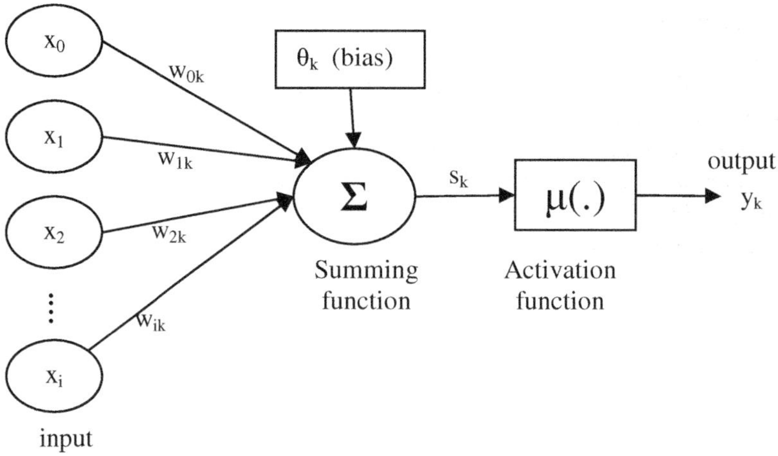

Fig. 7 The process of computing the output of a unit

Note that values of units are arbitrary but they should range from 0 to 1 (sometimes -1 to 1 range). In general, every unit has following aspects:

- A set of inputs connects to it. Each connection is defined by a weight
- Its weighted sum is computed by summing up all the inputs modified by their respective weights
- A bias value is added to weighted sum. This weighted sum is also called activation value.
- Its output is the outcome of activation function on weighted sum. Activation function is crucial factor in neural network.

Activation function is the squashing function which "squashes" a large weighted sum into possible smaller values ranging from 0 to 1 (sometimes -1 to 1 range). There are three types of activation function:

- *Threshold function* takes on value 0 if weighted sum is less than 0 and otherwise. So $\mu(x) = \begin{cases} 0 \; if \; x < 0 \\ 1 \; if \; x > 0 \end{cases}$
- *Piecewise-linear function* takes on values according to amplification factor in a certain region of linear operation. So $\mu(x) = \begin{cases} 0 \; if \; x \leq -\frac{1}{2} \\ x \; if \; x = -\frac{1}{2} < x < \frac{1}{2} \\ 1 \; if \; x \geq \frac{1}{2} \end{cases}$
- *Sigmoid function* takes on values in range [0, 1] or [-1, 1]. The formula of sigmoid function which is $\mu(x) = \dfrac{1}{1+e^{-x}}$.

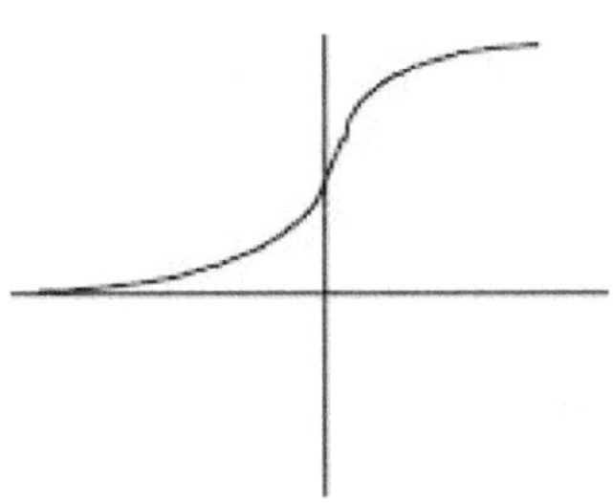

Fig. 8 Sigmoid function

There are two topologies of neural network:

- *Feed-forward neural network.* It is directed acyclic graphic in which flow of signal from input units to output units is one-way flow so-called feed-forward. There are no feedback connections

- *Recurrent neural network*. The graph contains cycles; so there are feedback connections in network.

It is necessary to evolve neural network by modifying the weights of connections so that they become more accurate. In other words, such weights should not be fixed by experts. The neural network should be trained by feeding it teaching patterns and letting it change its weights. This is learning process. There are three types of learning methods:

- *Supervised learning*. The network is trained by providing it with input and matching output patterns. These patterns are known as classes.
- *Unsupervised learning*. The output is trained to respond to clusters of pattern within the input. There is no a priori set of categories into which the patterns are to be classified.
- *Reinforcement learning*. The learning machine does some action on the environment and gets a feedback response from the environment. The learning system grades its action rewarding or punishable based on the environmental response and accordingly adjusts weights. Weight adjustment is continued until there is no change in weights.

Reinforcement learning is the intermediate form between supervised learning and unsupervised learning. We apply neural network into classifying corpus and such supervised learning algorithm used in this chapter is back-propagation algorithm.

4.2 Back-Propagation Algorithm for Classification

The back-propagation algorithm [4] is a famous supervised learning algorithm for classification, which is used in feed-forward neural network. It processes iteratively data tuples in training corpus and compares the network's prediction for each tuple to the actual class of the tuple. For each time it feeds a training tuple, the weights are modified in order to minimize the mean squared error between network's prediction and actual class. The modifications are made in backward direction, from output layer through hidden layer down to hidden layer. Back-propagation algorithm includes following steps:

- *Initializing the weights*: the weights are initialized as random real number which should be in space [0, 1]. Each bias associated to each unit is also initialized.
- *Propagating the input value forward*. Training tuple is fed to input layer. Given the unit j, if unit j is input unit, its input value denoted I_j and its output value denoted O_j are the same.

  $$O_j = I_j$$

  Otherwise if unit j is hidden unit or output unit, its input value I_j is the weighted sum of all output values of units from previous layer. The bias is also added to this weighted sum

  $$I_j = \sum_i w_{ij} O_i + \theta_j$$

Where w_{ij} is the weight of connection from unit i in the previous layer to unit j, O_i is the output value of unit i from the previous layer and θ_j is the bias of unit j.

The output value of hidden unit or output unit O_j is computed by applying activation function to its input value (weighted sum). Suppose activation function is sigmoid function. We have:

$$O_j = \frac{1}{1+e^{-I_j}}$$

- *Propagating the error backward*: The error is propagated backward by updating the weights and biases to reflect the error of network's prediction. Given unit j, if unit j is output unit then its error is computed as below:

  $$Err_j = O_j(1 - O_j)(T_j - O_j)$$

  If unit j is hidden unit, the weighted sum of the errors of the units connected to it in the next higher layer is considered when its error is computed. So the error of hidden unit is computed as below:

  $$Err_j = O_j(1 - O_j)\sum_k Err_k w_{jk}$$

  Where w_{jk} is the weight of the connection from unit j to a unit k in the next higher layer and Err_k is the error of unit k.

- *Updating the weights and biases* is based on the errors. The weights are updated so as to minimize the errors. Given Δw_{ij} is the change in weight w_{ij}, the weight w_{ij} is updated as below:

  $$\Delta w_{ij} = (l)\mathrm{Err_j O_i}$$
  $$w_{ij} = w_{ij} + \Delta w_{ij}$$

  Where l is learning rate ranging from 0 to 1. Learning rate helps to avoid getting stuck at a local minimum in decision space and helps to approach to a global minimum.

  The bias θ_j of unit j is updated as below:

  $$\Delta\theta_j = (l)\mathrm{Err_j}$$
  $$\theta_j = \theta_j + \Delta\theta_j$$

- *Terminating condition*: There are following terminating conditions:
  - All Δw_{ij} in some iteration are smaller than given threshold
  - Iterating through all possible training tuples.

4.3 Applying Neural Network into Document Classification

Given a set of classes **C** = {*computer science, math*}, a set of terms **T** = {*computer, programming language, algorithm, derivative*}. Suppose all input variables (units) are binary or Boolean, every document is represented as a set of input variables. Each term is mapped to an input variable in which value *1* indicates the existence of this term in document and otherwise value 0 indicates the lack of this term in document. So the input layer consists of four input units: *"computer"*, *"programming language"*, *"algorithm"* and *"derivative"*.

The hidden layer is constituted of two hidden units: *"computer science"*, *"math"*. These units (variables) are also binary or Boolean. The output layer has only one unit named *"document class"* which is binary or Boolean (0 – documents belong to *computer science* class and 1 – documents belong to *math* class). The evaluation function used in network is sigmoid function. Our topology is feed-forward neural network (showed in figure 4) in which the weights can be initialized arbitrarily.

Note that we denote Boolean value as *0* and *1* (instead of *true* and *false*) for convenience when representing neural network which only accepts numeric value for units.

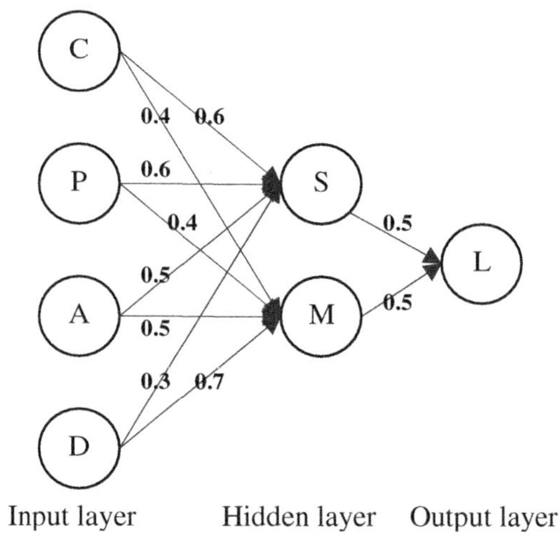

Fig. 9 The neural network for document classification

Note that *C, P, A* and *D* denote *"computer"*, *"programming language"*, *"algorithm"* and *"derivative"* respectively. *S* and *M* denote *"computer science"* and *"math"* respectively. *L* denotes *"doc class"*.

Given corpus **D** = {*doc1.txt, doc2.txt, doc3.txt, doc4.txt, doc5.txt*}. The training data is showed in following table in which cell (*i, j*) indicates the number of times that term *j* (column *j*) occurs in document *i* (row *i*).

Table 7 Term frequencies of documents

	computer	*programming language*	*algorithm*	*derivative*	***class***
doc1.txt	5	3	1	1	computer
doc2.txt	5	5	40	5	math
doc3.txt	20	5	20	55	math
doc4.txt	20	55	5	20	computer
doc5.txt	15	15	4	0.3	math
doc6.txt	35	10	45	10	computer

Table 8 Normalized term frequencies

	computer	*programming language*	*algorithm*	*derivative*	**class**
doc1.txt	0.5	0.3	0.1	0.1	computer
doc2.txt	0.05	0.05	0.4	0.5	math
doc3.txt	0.2	0.05	0.2	0.55	math
doc4.txt	0.2	0.55	0.05	0.2	computer
doc5.txt	0.15	0.15	0.4	0.3	math
doc6.txt	0.35	0.1	0.45	0.1	computer

Given threshold $\alpha = 0.4$, if the frequency of a term j in document i is equal or greater than α, we consider that term j exists in document i. Otherwise there is no existence of term j in document i. So each document is represented as a Boolean vector. Each element in such vector has two values: *0* and *1* (0 – the respective term occurs in document and 1 – otherwise). So each Boolean vector is the manifest of the occurrences of terms in a document. Corpus **D** becomes a set of Boolean vectors.

Table 9 Boolean document vectors

	computer	*programming language*	*algorithm*	*derivative*	**class**
doc1.txt	1	0	0	0	computer
doc2.txt	0	0	1	1	math
doc3.txt	0	0	0	1	math
doc4.txt	0.2	1	0	0	computer
doc5.txt	0.15	0	1	0	math
doc6.txt	0.35	0	1	0	computer

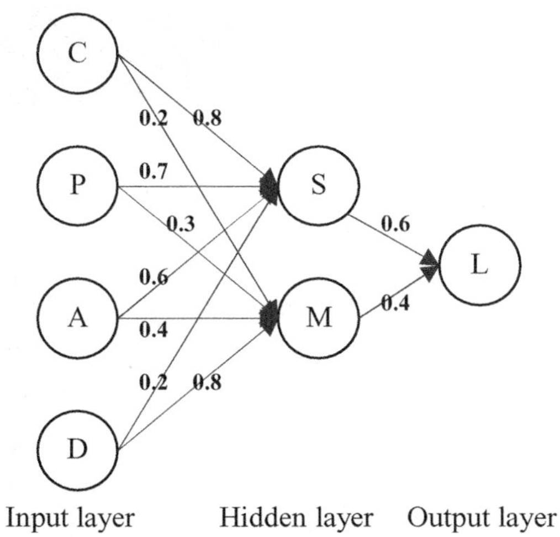

Fig. 10 Trained neural network

Such vectors are fed to our neural network in figure 3 for supervised learning. Back-propagation algorithm is used to train network; thus Boolean document vectors are considered training tuples. We suppose that the weights in neural network are changed as in figure III.3.9 after training process.

5 Discovering User Interests Based on Document Classification

Suppose in some library or website, user *U* do his search for his interesting books, documents… There is demand of discovering his interests so that such library or website can provide adaptive documents to him whenever he visits in the next time. This is adaptation process in which system tailors documents to each individual. Given there is a set of key words or terms {*computer, programming language, algorithm, derivative*} that user *U* often looking for, the his searching history is showed in following table:

Table 10 User's searching history

Date	*Keywords (terms) searched*
Aug 28 10:20:01	computer, programming language, algorithm, derivative
Aug 28 13:00:00	computer, programming language, derivative, algorithm
Aug 29 8:15:01	computer
Aug 30 8:15:06	computer

This history is considered as training dataset for mining maximum frequent itemsets. The keywords are now considered items. A itemset is constituted of some items. The support of itemset *x* is defined as the fraction of total transaction which containing *x*. Given support threshold *min_sup*, the itemset *x* is called *frequent itemset* if its support satisfies the support threshold (≥*min_sup*). Moreover x is *maximum frequent itemset* if *x* is frequent itemset and all super-itemsets of x are not frequent. Note that *y* is super-itemset of x if $x \subset y$. The itemset that has *k* items is called *k-itemset*. Tabel 9 shows the supports of *1*-itemsets.

Table 11 1-itemsets

1-itemset	*support*
computer	4
programming language	2
algorithm	2
derivative	2

Applying algorithm Apriori, it is easy to find maximum frequent itemsets given *min_sup* = 2. The maximum frequent itemset that user searches are showed in below table:

Table 12 The maximum frequent itemset that user searches

N_o	*itemset*
1	computer, programming language, algorithm, derivative

We propose the new point of view: "**The maximum frequent itemsets are considered as documents and the classes of such documents are considered as user interests**". Such documents may be called interesting documents. Which classes such interesting documents belong to are user interests. It means that discovering user's interests involves in classifying interesting documents. Suppose we have a set of classes **C** = {*computer science, math*}, a set of terms **T** = {*computer, programming language, algorithm, derivative*} and the set of classification rules in table 6. Each maximum frequent itemset that user searches is modeled as a document vector (so-called interesting document vector or user interest vector) whose elements are the support of its member items. Note that the supports of such items are showed in table 8.

Table 13 Interesting document vector

N_o	*vector*
1	(computer=4, programming language=2, algorithm=2, derivative=2)

Table 14 Interesting document vector is normalized

N_o	*vector*
1	(computer=*0.4*, programming language=*0.2*, algorithm=*0.2*, derivative=*0.2*)

Table 15 Nominal interesting document vector

N_o	*vector*
1	(computer=*medium*, programming language=*medium*, algorithm=*medium*, derivative=*medium*)

It is possible to use SVM or decision tree or neural network to classify documents. Hence we use decision tree as sample classifier for convenience because we intend to re-use classification rules in section III. Otherwise we must determine the weight vector W^* if applying SVM approach. However SVM approach is more powerful than decision tree with regard to document classification in case of huge training data.

Applying classification rule *2*, the interesting document belongs to class *compute science* because the frequency of "derivative" and "computer" are *medium* and *medium*, respectively. So we can state that user *U* has only one interest: *computer science*.

Note that in case of using neural network for document classification, interesting document vector is specified as Boolean document vector (or Boolean user vector). For example, given threshold $\alpha = 0.4$, if the frequency of a term j in document i is equal or greater than α, we consider that term j exists in document i. Otherwise there is no existence of term j in document i. We have a Boolean vector. So user U is modeled as a Boolean document vector. Such vector is also called Boolean user vector: $\boldsymbol{U} = \mathbf{(1, 0, 0, 0)}$. According to table III.3.19, we have

Table 16 Boolean document vector (or Boolean user vector)

Term	*Existence*
computer	1
programming language	0
algorithm	0
derivative	0

The Boolean user vector is considered as a document and the classes of such document are considered as user interests". Now document (user Boolean vector) U becomes a data tuple which is fed to trained neural network in figure III.3.10. It is easy to know the class of document U by checking the value of output unit in trained neural network. Suppose such output value is *0*, we can infer that document U belongs to class "*computer science*". So the interest of user U is "*computer science*".

6 Evaluation

Our approach includes following steps:

- Documents are represented as vectors
- Classifying documents by using decision tree or support vector machine or neural network
- Mining user's access history to find maximum frequent itemsets. Each itemset is considered an interesting document
- Applying classifiers into interesting documents in order to find their suitable classes. These classes are user interests.

Two new points of view are inferred from these steps:

- The series of user access in his/her history are modeled as documents. So user is referred indirectly to as document.
- User interests are classes that such documents are belong to

The technique of constructing vector model for representing document is not important to this approach. There are some algorithms of text segmentation for specifying all terms in documents. From this, it is easy to build up document vectors by computing term frequency and inverse document frequency. However the concerned techniques of document classification such as SVM, decision tree and neural network influence extremely on this approach. SVM and neural network is more effective than decision tree in case of huge training data set but it is not convenient for applying classifiers (weight vector W^*) into determining the classes of documents. Otherwise it is easy to use classification rules taken out from decision tree for this task.

References

1. Cristianini, N., Shawe-Taylor, J.: An Introduction to Support Vector Machines. Cambridge University Press, Cambridge (2000)
2. Mitchell, T.: Machine Learning. McGraw-Hill International, New York (1997)
3. Cortes, C., Vapnik, V.: Support vector networks. Machine Learning 20, 273–297 (1995)
4. Alrifai, M., Dolog, P., Nejdl, W.: Learner Profile Management for Collaborating Adaptive eLearning Applications. In: APS 2006: Joint International Workshop on Adaptivity, Personalization and the Semantic Web at the 17th ACM Hypertext 2006 conference, Odense, Denmark (August 2006)
5. Papatheodorou, C.: Machine Learning in User Modeling. In: Paliouras, G., Karkaletsis, V., Spyropoulos, C.D. (eds.) ACAI 1999. LNCS (LNAI), vol. 2049, pp. 286–294. Springer, Heidelberg (2001)
6. Rojas, R.: Neural Networks: A Systematic Introduction. Springer, Berlin (1996)
7. Lippmann, R.P.: An introduction to computing with neural nets. IEEE Transactions on Acoustics, Speech, and Signal Processing 1987 (1987)
8. Papatheodorou, C.: Machine Learning in User Modeling. In: Paliouras, G., Karkaletsis, V., Spyropoulos, C.D. (eds.) ACAI 1999. LNCS (LNAI), vol. 2049, pp. 286–294. Springer, Heidelberg (2001)
9. Han, J., Kamber, M.: Data Mining: Concepts and Techniques, 2nd edn. Elsevier Inc., Amsterdam (2006)

Network Analysis of Opto-Electronics Industry Cluster: A Case of Taiwan

Ting-Lin LEE

Abstract. The research scope in this study includes optoelectronics industry, especially LCD firms, and other complementary institutions in STSP. The purposes of this study are mainly to describe supply chain relationships networks of opto-electronics industry in STSP as fully as possible, tease out the prominent patterns in such networks, and discover what effects these relationships and networks have on organizations performance. In this regard, this research was conducted into two stages. In the first stage, some hypotheses (if some variables, for example network competence, network position and absorptive capacity, have significant effects on the innovation performance) will be verified through SPSS. We also expect to examine if the "network position" and "absorptive capacity" to the innovation performance has a meditative effect or not, and if they have significant influences on innovation performance. The second stage we adopted SNA to analyse further how the "network position" (such as direct ties, closeness centrality, betweenness centrality, coreness, and structural holes etc.) affect innovation performance.

In this study, most hypothesizes have been proved positively, and we further figure out the variable of "network position" exists meditative effect between network capacity and innovation performance. Simultaneously, network position also acts as mediator between network capacity and absorptive capacity. The results of this study contribute to a better understanding of how firms can utilize network benefits to enhance their innovation performance. Furthermore, "coreness centrality" is the most interpretable position variable for innovation performance.

Keywords: social network analysis, network competence, absorptive capacity, Southern Taiwan Science Park (STSP).

Ting-Lin LEE
Department of Asia Pacific Industrial and Business Management,
National University of Kaohsiung 700,
Kaohsiung University Road, Kaohsiung 811, Taiwan
e-mail: linda_lee@nuk.edu.tw

I.-H. Ting et al. (Eds.): Mining and Analyzing Social Networks, SCI 288, pp. 161–182.
springerlink.com

1 Introduction

Since the IC industry in Taiwan developing maturely into the 21st century, next emerging high-tech industry in Taiwan has already come around the corner that is "opto-electronics industry." Thanks to governments' policy planning, "North (Shin-Chu Science Park, SCSP) IC, South (Southern Taiwan Science Park, STSP) Opto-electronics," engines in the high-tech industry have been fuelled by two major focuses and each of them locates in different Parks. Obviously, each industry has developed an industry network along with its cluster in the Park. The STSP, for example in 2006, is homed by more than 38 opto-electronics corporations which were has been approved by National Science Council, and 29 of them have production. The total revenue in the STSP has already topped $4,516 hundred million in 2006. Opto-electronics industry, particularly, has already taken above 71% in the STSP. Sales revenue for opto-electronics industry was NT$3,224 million in 2006, an increase of 16 times over 2001. Corporations in the industry has benefited from the policy, and part of economic gains coming from the cluster in the Park. Obviously, the STSP has turned out to be an important cluster of opto-electronics in Taiwan.

Science Parks have been linked to the economic development of countries and the creation of local competitive advantages. According to Hakansson and Ford (2002) point out that there are three paradoxes (e.g. opportunities and limitation in networks; influence and being influenced in a network; controlling and being control in networks) that can be detected in a network. Nevertheless, business cluster is one of the formats of network. This study examine if business clusters (such as science parks) faced these paradoxes. These three paradoxes give us insights about there are insistence advantage and risk surrounding the network. Porter (2000) argues that location as a single factor is losing its competitive advantage. However, as firms join clusters, the meaning of location in the sense of firm agglomeration becomes important again. Government set many special areas to attract foreign investments. Companies also want to gain benefits from Science Park by means of policy subsidy or cooperation with other firms. The main motivation of this research is to see if firms can really get the resources they want from the cluster. In order to understand relationships among optic-electronic firms, especially in LCD firms, in the STSP, Social Network Analysis (SNA) approach and related statistics will be used in the research.

According to statements above, we are interested in the following issues related to the firms in the Southern Taiwan Science Park: 1) the communication patterns among LCD firms in the STSP and find out the key players in LCD industry, 2) the relationships between network competency, network position, absorptive capacity and innovation performance, 3) the effect of network features, such as degree, betweenness, closeness, coreness, and structural holes etc., on the innovation performance.

2 Literature Review

2.1 Network Competence

Early researches focus on competence in terms of firm's ability to satisfy customer and these issues were related to relational marketing. But, recently researcher attach importance to relationships between firms' (e.g. Ritter and Gemunden 2004; Gemunden, Ritter, and Heydebreck, 1996; Ritter and Gemunden, 2003; Ritter, Wilkinson, and Johnston, 2002). Ritter and Gemunden (1997) introduced the network competence as firm's specific characteristic. Ritter (1999) claims this ability as "network competence" and defines it "... the degree of network management task execution and the degree of network management qualification possessed by the people handling a company's relationships" (Ritter 1999, p.471; Ritter et al., 2002). According to its definition, network competence enables firm to establish relations and manage their relationships with multiple partner companies. In other words, the ability to cultivate network competence can help firm establish and utilize relationships with others organizations. The more competence a firm has with respect to establishing relationships, the more likely the company will find itself embedded in a rich network of relationships. Hence, we hypothesize:

H1: There is a significant relationship between a company's level of network competence and its network position.

Many researchers have examined relationship between network competence and innovation performance, and they descript the advantage of higher degree of network competence. Firms with higher degree of network competence can efficiently establish relationship with other firms and promote the communication of information to contribute innovation performance (Biemans, 1992; Gemunden et al., 1996). Furthermore, Scholars argue that if firms have higher network competence, then they can obtain market-oriented path to innovation success (Ritter & Gemunden, 2004). Therefore, it helps firms set up better market strategy to introduce new innovative product. Besides, Li and Calantone (1998) assumed firms with higher network competence also have higher market knowledge competence, so it can lead to innovation success. Some researchers have proposed a link between a company's network competence and its innovation success (Biemans, 1992; Gemunden et al., 1996; Heydebreck, 1996; Ritter & Gemunden, 2004). Here, we also want to examine network competence and innovation performance in Taiwan's Optoelectronics industry. Hence we hypothesize as follows:

H2: There is a significant relationship between a company's level of network competence and its innovation performance.

2.2 Absorptive Capacity

There are several scholars point out that absorptive capacity as an innovation platform which can increase the speed and frequency of incremental, it primarily attribute on the firm's existing knowledge base (Kim & Kogut, 1996; Anderson & Tushman, 1990; Helfat, 1997). Cohen and Levinthal (1990) argued that absorptive capacity with valuable input factor can increase and facilitate a firm's innovation to the R&D process. Still, Hurry, Miller and Browman (1992) noted that this process is 'self-reinforcing' which means that innovation and absorptive capacity can reinforce each other. Certainly the innovative output will be turned into the external knowledge which will be a resource for other companies after a period. We focus on and study the relation between the results of the absorptive capacity and innovation performance mainly. In our research, absorptive capacity is defined as a firm's learning capability. Also Tsai (2001), Lane & Klavans (2005) has examined the relation between absorptive capacity and innovation performance. Fallowing this clue, in our research we bring into absorptive capacity as mediating variable. We therefore hypothesize as follows:

H3: There is a significant relationship between a company's level of network competence and its absorptive capacity.

H4: There is a significant relationship between a company's level of absorptive capacity and its innovation performance.

2.3 Network Position

There are some evidences shows positive relationship between network position and innovation from a network perspective (Tsai, 2001). Tsai (2001) points out that in an intraorganizational network a higher centrality is significantly and positively related to innovation. This finding may also apply in an interorganizational situation. Network position is a descriptor of ''social structure'' and play important role in networks (Coleman et al., 1990; Tsai & Ghoshal, 1998). Position can enhance actor's capability to develop innovative or creative value to achieve specific goals. Different network positions allow different levels of access to information or knowledge. We therefore hypothesize as follows:

H5: There is a significant relationship between a company's network position and its absorptive capacity.

Information and knowledge can viewed as external resources. Central network positions offer more opportunities to access these resources than peripheral positions. These resources represent "fuel" in the innovation process, driving innovation performance (Tsai, 2001). We therefore hypothesize as follows:

H6: There is a significant relationship between a company's network position and its innovation performance.

For further test which network position to the last influences innovation most, some variables were selected to observe.

2.3.1 Direct Ties and Innovation Performance

Direct ties could affect a firm's innovative output positively (Ahuja, 2000), and this is because direct ties provide three substantive benefits: First, direct ties facilitate knowledge sharing (Berg, Duncan, & Friedman, 1982). When collaboration relationship exists among firms, the relative industry knowledge is available to all partners. Thus, each partner can potentially receive a greater amount of knowledge from a collaborative firm. Second, collaboration facilitates bringing together complementary skills from different firms (Arora & Gambardella, 1990). Technology often demands the simultaneous use of different sets of skills and knowledge based in the innovation process (Arora & Gambardella, 1990; Powell, Koput, & Smith-Doerr, 1996). Hence, this different knowledge from variety of firms could facilitate innovative ideas. Third, the effect of direct ties emerges through scale economies in research that arise when larger projects generate significantly more knowledge than smaller projects. Collaboration enables firms to take advantage of such scale economies. In addition, in biotechnology start-ups, Shan, Walker, and Kogut (1994) found that the greater the number of collaborative linkages formed by a start-up, the higher the innovation performance. And Ahuja (2000) found the same results in the chemical industry. Besides, firm's communication relationship with knowledge producing institutions, such as universities, research institutes, technology-providing firms and bridging institutions such as providers of technical or consultancy services all enable knowledge sharing and exchange (Drejer, Kristensen, & Laursen, 1999). Furthermore, firms without collaboration relationship maybe also have communication relationship. Communication relationship makes firms to exchange information about technology, industrial development and etc. What information do these firms exchange depends on their ties strength. Hence, the more direct ties a firm maintains, the more information the firm holds. Then firms may exploit the information to their advantage (Burt, 1992). Accordingly, we hypothesize that:

H6a: The more direct ties a firm has, the greater the firm's innovation performance will be.

2.3.2 Closeness Centrality and Innovation Performance

An interfirm linkage can be a channel of communication between the firm and many indirect contacts (Mizruchi, 1989; Davis, 1991; Haunschild, 1993; Gulati, 1995). A firm's partners bring the knowledge and experience from their interactions with their other partners to their interaction with the focal firm, and vice versa (Gulati & Garguilo, 1999). A firm's linkages therefore provide it with access not just to the knowledge held by its partners but also to the knowledge held by its partner's partners (Gulati & Garguilo, 1999). The network of interfirm linkages thus serves as an information conduit, with each firm connected to the network being both a recipient and a transmitter of information (Rogers & Kincaid, 1981). Hence, if a firm could reach any other firm in the shortest distances, namely, higher closeness centrality, the firm could get knowledge and information in the shortest time (Freeman, 1979). Other things being equal, firms that spend less time to obtain knowledge and information, are likely to have effect on innovation performance than firms who spent more time. Thus, we hypothesize:

H6b: There is a positive relationship between a firm's closeness centrality and its innovation performance.

2.3.3 Betweenness Centrality and Innovation Performance

Betweenness is an indicator of network centrality based on Freeman's (1979) measure of betweenness, which captures the extent to which firms sit astride network pathways between other organizations. Betweenness centrality indicates a firm's ability to control information flows within communication network. Hence, a firm controls more information flow, it could gather more important information. Thus, the firm could perform better than others in innovation process. Similarly, Smith-Owen and Powell (2004) proved the effect of betweenness centrality on innovation performance in Boston biotechnology community. Accordingly, we hypothesize that:

H6c: There is a positive relationship between a firm's betweenness centrality and its innovation performance.

2.3.4 Coreness and Innovation Performance

The core/periphery models developed by Borgatti and Everett (1999) provide a useful analytical tool that represents the classic idea of a core formed by a group of densely connected actors, in contrast to a more loosely connected class of actors making up the periphery of the system. In other words, a network has a core/periphery structure if the network can be partitioned into two sets: a core whose members are densely tied to each other and a periphery whose members have more ties to core members than to each other. From this viewpoint, firms in core have more dense communication than firms in periphery and thus transfer more information and knowledge among them. Accordingly, firms in core have better innovation performance than firms in periphery. In this research, we use coreness as an indicator of core/periphery which was developed by Borgatti and Everett (1999). The higher the coreness of a firm, the better the firm's innovation performance will be. Thus, it can be stated that:

H6d: There is a positive relationship between a firm's coreness and its innovation performance.

2.3.5 Structural Holes and Innovation Performance

Recent research suggests that a firm's ego network is likely to be important to innovation such as the extent of connectivity between a firm's partners (Burt, 1992). The underlying mechanism posited by Burt is that actors in a network rich in structural holes will be able to access novel information from remote parts of the network, and exploit that information to their advantage (Burt, 1992). Moreover, a structural hole indicates that the actor on either side of the hole have access to different flows of information (Hargadon & Sutton, 1997). Many structural holes in ego's network will increase ego's access to diverse information and, thus, enhance innovation performance (Ahuja, 2000). According, we hypothesize that:

H6e: The greater the structural holes bridged by a firm, the greater the firm's innovation performance.

3 Methodology

3.1 *Questionnaire Design*

In our research, the content of the questionnaire is divided into four major parts. In the first part we measure the degree of firm's absorptive capacity, including three sub-constructs: knowledge acquisition, knowledge diffusion, and knowledge utilization. Part two captures the communication network between firms in STSP. On a list showed all organizations of the optoelectronics industry in STSP, the respondents were asked to indicate what kind of relationship with other firms and institutions they have. There are three types of relationship proposed in the questionnaire: buyer/supplier, competitor, and other. In other words, the analytic network in this research consists of three kinds of communication relationships, and three groups of actors. In the third part the respondents were asked to evaluate their firm's level of network competence. We use three items based on the work of Abernathy & Utterback (1988) - efficiency of R&D process, number of successful new product developments, and time-to-market for newly developed products, to measure a firm's innovation performance. We finally aggregate these three items into an index of innovation performance. The final part includes basic company information.

3.2 *Data Collection and Analysis Structure*

The STSP gathers optoelectronics manufactures and constructs a complete social network. It can be seen that a complete supply chain, especially for LCD firms, was formed in STSP and became more worthy as an observation target for our research. Three copies of Questionnaire were sent to each firms of the optoelectronics cluster in STSP, the respondents include senior managers, R&D staffs, and purchasing/sales agents at different levels/department. By the limited sample size, sampling methods have not been taken into account. Data collection was conducted from January till May 2007. Altogether, 43 firms and institutions were contacted. 27 questionnaires were retrieved, which is a return rate of 62.7%. If we only consider LCD firms, which is the core of our first stage research, the return rate increases up to 85.7% (refer to Table 1).

Table 1 The statistics of Questionnaire copies

		Types of Actors in STSP			
		LCD firms	Institutions	Other Firms	Total
Copies distributed		21	10	12	43
Missing/invalid copies		3	6	7	16
Copies returned		18	4	5	27
Response rate		85.7%	40%	41.6%	62.7%
Position	Manager numbers	16	4	3	23
	Staff numbers	2	0	2	4

The first stage in our research mainly focuses on the LCD firms' communication network; the second stage, the network boundary will be extended to institutions in STSP and even to optoelectronics firms of STSP. The purposes of this extension are to see if firms' network position variable changed with extended network boundary, and see if the changed network position variables' effect on innovation performance. Accordingly, we design four kinds of communication network based on three kinds of relationships and three groups of actors. They are LCD network (LCDNet), LCD network with institutions (LCD-InstNet), collaborative network (ColNet), and the whole optoelectronics network in STSP (OptoNet). In a word, LCDNet contains only LCD firms and their vertical collaborative relationships. LCD-InstNet contains LCD firms and related institutions, and the vertical and assistant relationships among them. ColNet contains all actors and relationships except horizontal competitor relationship. OptoNet contains all actors which include cooperators, competitors, and institutions and theirs relationships.

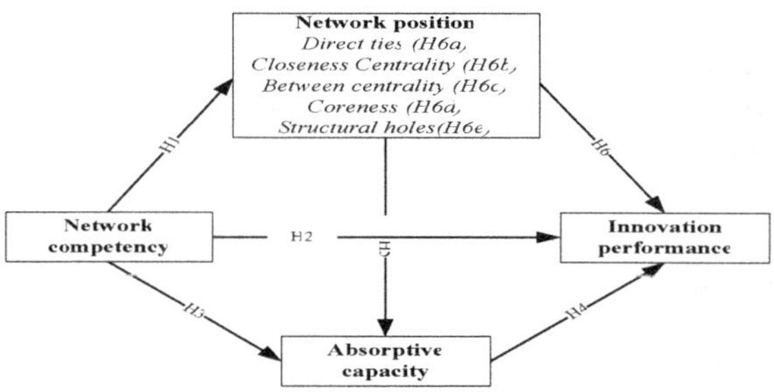

Fig. 1 Research structure

4 Empirical Analysis and Discussion

4.1 Hypotheses Prove for First Stage

In this section we mainly discuss our hypothesis 1 to 6, and conduct testing. Regression analysis are mainly based on the linear relation of valuables, and go a step further predict relationships between variables. In Table 2, we can figure the relations among those hypothesis variables, and the means and sum of square of each variable. Moreover, Table 3 shows the correlation between variables and significant differences.

4.1.1 Network Competence Related Hypothesis Proof (H1~H3)

The influence of a company's network competence on its network position, innovation performance, absorptive capacity as shown in Table 4. The results of the regression analysis show that there is respectively a significant and positive correlation between them. If a company possesses higher networking capacity, it would positively influence its position in the network because networking competence is connected with other partnership or level of communication.

Table 2 Mean and standard deviation of all research variables

Hypothesis	Variable format	Variable name	Mean	Standard deviation	N
H1	Independent Variable	Network competence	69.083	5.392	18
	Dependent Variable	Network position	0.2077	0.110	
H2	Independent Variable	Network competence	69.083	5.392	18
	Dependent Variable	Innovation performance	14.278	2.492	
H3	Independent Variable	Network competence	69.083	5.392	18
	Dependent Variable	Absorptive capacity	68.584	6.989	
H4	Independent Variable	Network position	0.2077	0.110	18
	Dependent Variable	Absorptive capacity	68.584	6.989	
H5	Independent Variable	Network position	0.2077	0.110	18
	Dependent Variable	Innovation performance	14.278	2.492	
H6	Independent Variable	Absorptive capacity	68.584	6.989	18
	Dependent Variable	Innovation performance	14.278	2.492	

Table 3 Correlation Matrix of variables

	Network competence	Absorptive capacity	Network position	Innovation performance
Network competence	1			
Absorptive capacity	0.706(**)	1		
Network position	0.482(*)	0.736(**)	1	
Innovation performance	0.661(**)	0.609(**)	0.621(**)	1

Whenever a company embraces higher networking capacity, it simultaneously implies the close interaction with other partners and stands on the position near the network anchor.

The higher the communication level is the easier sharing information will be. This source of information could be come from a more market oriented down-stream firm or/and up-stream firm. This affluent information can bring firms to increase its efficiency and successfulness of innovation.

Besides, if a company possesses a high networking competence, which means the company contains a kind of capacity of making other firms "willing" and "must" to interact and exchange information. Once a company embraces the ability of handling the information and internalizes them into its own knowledge, and then becomes its competitive advantages. The transformation process is exactly like the procedure of absorptive capacity. Furthermore, the research results also prove that there are positive relations between the level of network competence and absorptive capacity.

4.1.2 Absorptive Capacity Related Hypothesis Proof (H4)

The influence of a company's absorptive capacity on its innovation performance as shown in Table 4. Obviously, the stronger absorptive capacity a company contains the higher innovative efficiency a company will be. The outcome is in line with what scholars had previously examined as well (e.g. Tsai, 2001; Lane and Klavans, 2005). The result does not come as a surprise: companies with better absorptive capacity can benefit from external knowledge acquisition, and internalize it to enhance internal capability. Therefore, the output of knowledge can directly correspond to a company's innovation performance.

4.1.3 Network Position Related Hypothesis Proof (H5~H6)

The results of the regression analysis indicate that it is respectively a significant and positive correlated between network position and absorptive capacity, innovation performance. The more central network position compared to surrounded position can easily receive messages and information, and it will influence the level of absorptive capacity and its innovation performance. Our outcome is in line with what scholars had examined as well (e.g. Tsai and Ghoshal, 1998). Our research finding once more emphasizes the importance of network position.

Table 4 Regression analysis of each model

Model	Unstandardized Coefficient(b)	Std. Beta(β)	t value	Sig.	R^2	Adjusted R^2
Network competence *(H1~H3)*			**Network position**			
	0.010	0.482	2.198	0. 043**	0.232	0.184
			Innovation Performance			
	0.305	**0.661**	3.519	0.003***	0.436	0.401
			Absorptive capacity			
	0.916	**0.706**	3.992	0.001***	0.499	0.468
Absorptive capacity *(H4)*			**Innovation performance**			
	0.217	0.609	3.072	0.007***	0.371	0.332
Network position *(H5~H6)*			**Absorptive capacity**			
	46.818	0.736	4.348	0.000***	0.542	0.513
			Innovation performance			
	14.079	0.621	3.166	0.006***	0.385	0.347

Note: *p<0.1, **p<0.05, ***p<0.01

4.1.4 Multiple-Regression Analysis

The using of multiple-regression is to examine *if network competence and network position have a significant effect on absorptive capacity.* According to Table 5, the results show that it is significant and positive relation between network competence and network position on absorptive capacity at the same time. Hence, it can be gone forward to examine if network position is a mediator variable. In H3, the Std. Beta of network competence is 0.706 (β=0.706). But in this Multiple-regression model the Std. Beta of network competence decrease to 0.458 (β=0.458). Therefore we can argue that there is a partial mediate effect between network competence and absorptive capacity.

Table 5 Multiple-regression analysis both network position and network competence on absorptive capacity and innovation performance

Model	Unstandardized Coefficient(b)	Std. Beta (β)	t value	Sig.	R^2	Adjusted R^2
			Absorptive capacity			
Network position	32.778	0.515	3.209	0.006***	0.703	0.663
Network competence	0.594	**0.458**	2.854	0.012**		
			Innovation performance			
Network position	8.934	0.394	2.005	0.063*	0.555	0.496
Network competence	0.218	**0.471**	2.397	0.030**		

Note: *p<0.1, **p<0.05, ***p<0.01

Next, we examine *if network competence and network position have significant effect on innovation performance.* The results also indicate that it is significant and positive related of network competence and network position on innovation performance. There is the same with above result. In H2, the Std. Beta of network competence is 0.661 (β=0.661). But in this Multiple-regression model the Std. Beta of network competence decrease to 0.471 (β=0.471). Although, the P-value of network position is 0.063(p=0.063), but we can say it still has partial significant. Therefore we can argue that network position has partial mediate effect between network competence and innovation performance.

Moreover, the authors want to examine *if network competence and absorptive capacity have significant effect on innovation performance.* The result shows that there are no significant relationships among the mentioned variables (refer Table 6). Finally, the study examine *if absorptive capacity and network position have significant effect on innovation performance.* The result still shows that there are no significant relationships among the mentioned variables (refer to Table 7). Hence, it is reasonable to doubt that maybe there is some moderate effect that unhuman manipulate on these variables. There might have several reasons to explain above results. First, data collected are incomplete. The ideal data retrieve rate is 100% in

Table 6 Multiple-regression analysis both network competence and absorptive capacity on innovation performance

Model	Unstandardized Coefficient(b)	Std. Beta (β)	t value	Sig.	R^2	Adjusted R^2
	Innovation performance					
Network competence	0.212	0.460	1.947	0.102	0.477	0.407
Absorptive capacity	0.101	0.284	-0.138	0.298		

Note: *p<0.1, **p<0.05, ***p<0.01

Table 7 Multiple-regression analysis coefficient both absorptive capacity and network position on innovation performance

Model	Unstandardized Coefficient(b)	Std. Beta (β)	t value	Sig.	R^2	Adjusted R^2
	Innovation performance					
Absorptive capacity	0.119	0.332	1.160	0.264	0.436	0.361
Network position	8.529	0.376	1.312	0.209		

Note: *p<0.1, **p<0.05, ***p<0.01

Table 8 The Summary of empirical results

Hypothesis	***Result***
H1: There is a significant relationship between a company's level of network competence and its network position.	*accepted*
H2: There is a significant relationship between a company's level of network competence and its innovation performance.	*accepted*
H3: There is a significant relationship between a company's level of network competence and firm's absorptive capacity.	*accepted*
H4: There is a significant relationship between a company's level of absorptive capacity and its innovation performance.	*accepted*
H5: There is a significant relationship between a company's network position and its absorptive capacity.	*accepted*
H6: There is a significant relationship between a company's network position and its innovation performance.	*accepted*

SNA study. In practice, it is difficult to achieve this goal. Because of some ignored firms' communication, it might result in insignificant. Second, maybe there still exist some moderate effects that were not found. Third, the sample is too small for multiple-regression. But we still believe these independent variables have significant affect to innovation performance.

4.2 Network Analysis

The following section, five hypotheses about network position variables (degree, closeness, betweenness, structural holes, coreness etc.) and innovation performance will be tested in detail. The results of correlation analysis are shown in Table 9. There are significantly positive correlations among independent variables. This is because our independent variables are all network position variables, although they capture different position concepts. However, high correlations among independent variables could lead to collinearity in multiple regressions. In addition, our research respectively focuses on the effect of different network position variables on innovation performance and seeks to examine which network position variable could explain innovation performance most. Hence, this section will adopt simple linear regression to test our hypotheses.

Table 9 Means, standard deviations, and correlations

	Mean	S.D	Degree	Closeness	Betweenness	Structural Holes	Coreness	Innovation Performance
Degree	11.44	6.35						
Closeness	0.55	0.09	0.975**					
Betweenness	0.06	0.09	0.917**	0.875**				
Structural Holes	0.63	0.20	0.718**	0.804**	0.550*			
Coreness	0.21	0.11	0.937**	0.942**	0.861**	0.671**		
Innovation Performance	14.28	2.49	0.486*	0.509*	0.487*	0.337	0.621**	

Four graphs represent network structures of optoelectronic industry in the STSP and in different scope. Code A represents LCD industry companies. Code B represents other optoelectronic industry companies which settle in the STSP and support/compete with Code A firms. Code C represents research institution units which settle in the STSP and cooperate with Code A firms.

Figure 2 shows the LCD industry in the STSP (LCDnet), which are our research main targets. The size of nodes is weighted by degree centrality. It can be observed A3 and A4 are key players in this network structure. Besides, the evidence indicates that the LCD cluster is a close cluster. The ties maintained by these two key players almost make up the whole LCD cluster, especially for firm A4 that almost has collaboration relationships with all other LCD firms. In a close network, through frequent social interaction, actors are likely to have shared values, shared norms and even collective behavior (Nooy, et al., 2005). Moreover, LCD industry is a cluster that places importance on supply chain's collaboration. Hence, the two key players would play a role of industrial leader to lead LCD firms to achieve a better overall industry performance.

Figure 3 shows the LCD companies plus collaborative research institutions (LCD-InsNet). Figure 3 is a visualization graph, where the green nodes (C1~C10) indicate institutions and red nodes indicate LCD firms (A1~A21). Institution C1 is the biggest node in the graph because it is the official administration of STSP. Hence, it must have most communication with firms in any industry in STSP' institutions. When the research institutions were included in this network, the position of A1 becomes even more central. A3 and A4 remain the key players in this network.

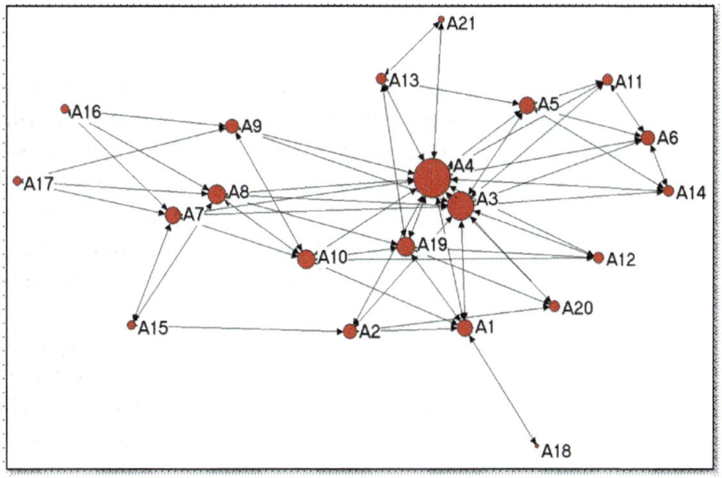

Fig. 2 Network structure LCDNet

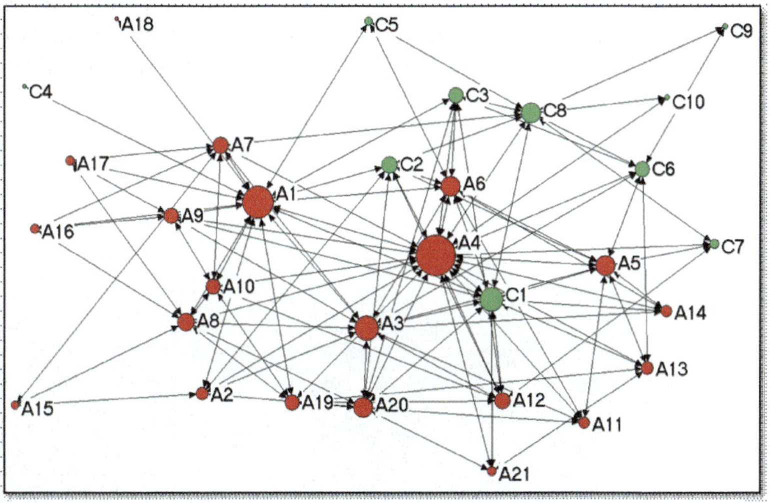

Fig. 3 Network structure of LCD-InsNet

In Figure 4 (ColNet), red nodes indicate LCD firms, green nodes indicate institutions, and purple nodes indicate other firms of optoelectronics industry. The sizes of nodes are weighted by degree centrality. From Figure 4, we can see that almost other firms of optoelectronics industry are in the periphery. This is because the direct ties of other firms of optoelectronics industry are smaller than others.

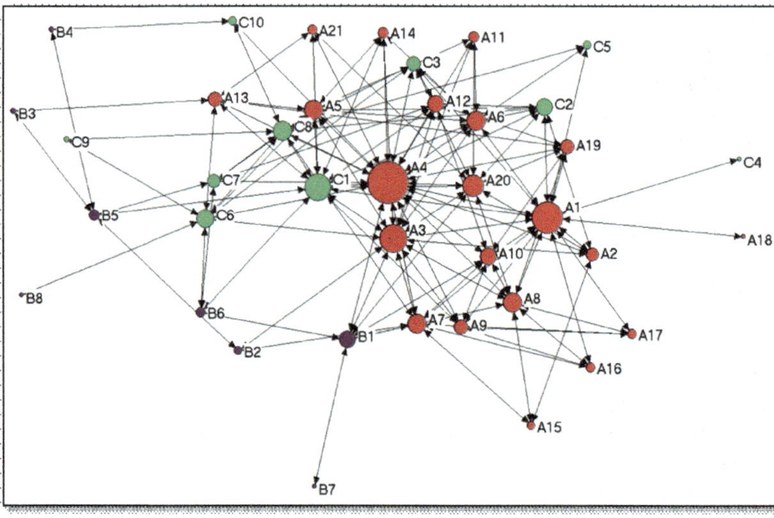

Fig. 4 Network structure of ColNet

Figure 5 shows the whole communication network among optoelectronic companies (B1~B8) and research institutions in the STSP. It includes collaborative relationships (supply chain), competitive relationships and horizontal relationships

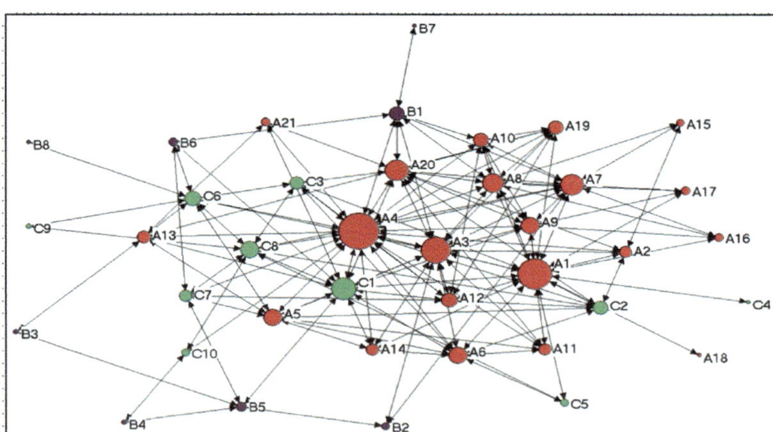

Fig. 5 Network structure of Optonet

(e.g. assistance relationship). The only difference between ColNet and OptoNet is the later contains communication relationships with competitors. In Figure 5, because communication relationships between competitors are quite less, the difference between Figure 4 and Figure 5 are also quite small.

4.3 Hypotheses Prove for the Second Stage

4.3.1 Direct Ties and Innovation Performance

The regressive results of H6a (Table 10) indicate that direct ties of four kinds of network have significantly positive effect on innovation performance. Hence, the more direct ties a firm maintains, the more information the firm holds. Then because of these information and knowledge from communication among actors, the firm has better innovation performance. Moreover, direct ties of LCD-InsNet have better explanation on innovation performance.

Table 10 Regression analysis of position variables on innovation performance of each communication network

Direct Ties	Innovation Performance					
	Unstandardized Coefficient(B)	Standardized Coefficients(ß)	t value	Sig,	R^2	Adjusted R^2
LCDNet	0.191	0.486	2.223	0.041**	0.236	0.188
LCD-InstNet	0.170	0.664	3.556	0.003***	0.441	0.407
ColNet	0.156	0.641	3.343	0.004***	0.411	0.374
OptoNet	0.126	0.577	2.824	0.012**	0.333	0.291
Closeness	Innovation Performance					
LCDNet	14.008	0.509	2.368	0.031**	0.260	0.213
LCD-InstNet	19.216	0.642	3.347	0.004***	0.412	0.375
ColNet	22.347	0.637	3.304	0.004***	0.406	0.368
OptoNet	19.323	0.570	2.778	0.013**	0.325	0.283
Betweenness	Innovation Performance					
LCDNet	13.155	0.487	2.231	0.040**	0.237	0.190
LCD-InstNet	18.889	0.572	2.788	0.013**	0.327	0.285
ColNet	20.507	0.549	2.625	0.018**	0.301	0.244
OptoNet	20.627	0.537	2.545	0.022**	0.288	0.244
Coreness	Innovation Performance					
LCDNet	14.079	0.621	3.166	0.006***	0.385	0.347
LCD-InstNet	18.532	0.703	3.950	0.001***	0.494	0.462
ColNet	18.566	0.697	3.888	0.001***	0.486	0.454
OptoNet	16.425	0.631	3.254	0.005***	0.398	0.361
Structural Holes	Innovation Performance					
LCDNet	4.159	0.337	1.432	0.171	0.114	0.058
LCD-InstNet	5.531	0.431	1.909	0.074*	0.186	0.135
ColNet	5.294	0.418	1.838	0.085*	0.174	0.123
OptoNet	5.747	0.441	1.965	0.067*	0.194	0.144

Note: *p<0.1, **p<0.05, ***p<0.01

Note: LCDNet: Communication Network

LCD-InstNet: LCD Communication Network with Institutions

ColNet: LCD collaborative network

OptoNet: the whole optoelectronics network with Institutions in STSP

4.3.2 Closeness Centrality and Innovation Performance

The regressive results of H6b indicate that closeness centrality of four kinds of network have significantly positive effect on innovation performance. Hence, it means that firms could reach other actors in shorter time and they get more information and knowledge faster. Therefore they can enhance innovation performance. Moreover, closeness centrality of LCD-InsNet has better explanation on innovation performance.

4.3.3 Betweenness Centrality and Innovation Performance

The regressive results of H6c indicate that betweenness centrality of four kinds of network have significantly positive effect on innovation performance. Hence, a firm controls more information flow, it could gather more important information. Then, the firm could perform better than others in innovation performance. Also, betweenness centrality of LCD-InsNet has better explanation on innovation performance.

4.3.4 Coreness and Innovation Performance

The regressive results of H6d indicate that coreness of four kinds of network have significantly positive effect on innovation performance. Hence, firms in core have better innovation performance than firms in periphery. Moreover, coreness of LCD-InsNet has better explanation on innovation performance.

4.3.5 Structural Holes and Innovation Performance

The regressive results of H6e indicate that structural holes of only three kinds of network have significantly positive effect on innovation performance. Hence, more structural holes in actor's network will increase actor's access to diverse information and, thus, have better innovation performance. Moreover, structural holes of OptoNet have better explanation on innovation performance.

Table 11 Summary of empirical results

Hypotheses	Results
H6a: The more **direct ties** a firm has, the greater the firm's innovation performance.	Supported
H6b: There is a positive relationship between a firm's **closeness centrality** and its innovation performance.	Supported
H6c: There is a positive relationship between a firm's **betweenness centrality** and its innovation performance.	Supported
H6d: There is a positive relationship between a firm's **coreness** and its innovation performance.	Supported
H6e: The greater the **structural holes** bridged by a firm, the greater the firm's innovation performance.	Partially Supported

4.4 Position Variables in Explaining Innovation Performance

From the empirical results, we can find that almost every network position variable had significantly positive effect on innovation performance, in spite of each position variables' adjusted R square on innovation performance are different, namely, each position variables' goodness of fit for innovation performance are different.

In the past, many scholars proved that network position had significant effect on firm's performance or innovation performance (Zaheer & Bell, 2005; Ahuja, 2000; Tsai, 2001). However, they often apply different network position variable to explain innovation performance in different industries. Hence, one objective of this research is to identify which network position variable is most suitable for explaining the network of LCD industry which emphasizes collaboration of supply chain in explaining innovation performance.

We use each adjusted R square (degree of explanation toward innovation performance) of regression analysis to present the distinction among different position variables and networks in Figure 6. Consequently, it shows that the explanation of coreness is the highest in each network. In other words, coreness offers the best explanation for innovation performance in all four networks. According to Borgatti and Everett (1999), in contrast to a more loosely connected class of actors, a core formed by a group of densely connected actors makes up the periphery of the network (refer to Figure 7). From the analysis results, it is likely that there are two key firms which dominate over other LCD firms. Important information and knowledge are mostly held and distributed by them, thus they have better innovation performance than others. In other words, the important information and knowledge often held by firms who are core in LCD industry. Hence, it can be concluded that when key firms in a network hold critical information, knowledge, or know-how, it is suitable to use coreness to explain innovation performance.

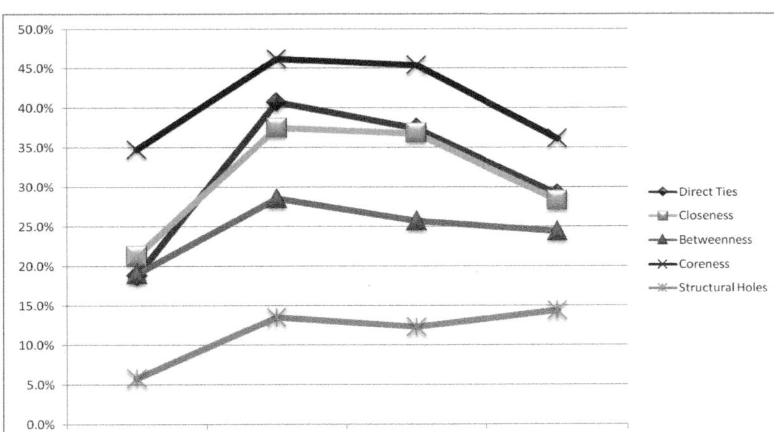

Fig. 6 Differences among position variables in explaining innovation performance

On the contrary, structural holes offer least explanation in each network. In addition, all network position variables get the best explanation toward innovation performance in LCD-InsNet. The explanatory power of position variables gradually decreases from LCE-InsNet to OptoNet, except for structural holes. It is likely that direct ties, closeness, betweenness, and coreness are best to be explained toward innovation performance in LCD-InsNet. Because this cluster contains all LCD firms and related institutions, the cluster is a fully complete network from an industrial cluster's perspective.

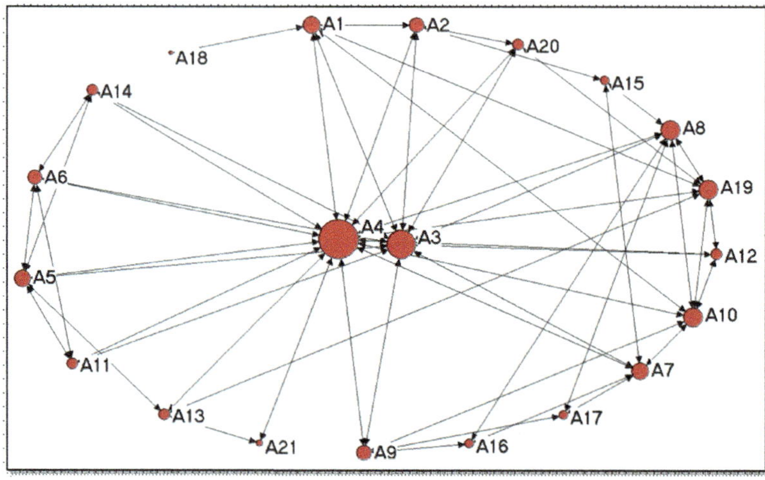

Fig. 7 Two key players in LCDNet

4.5 Remarks

All in all, network position indeed influences innovation performance from our results of analysis. We expected that a firm with more direct ties will lead him to access more information and knowledge and then get a better innovation performance. This hypothesis was supported. It means that rich and diverse sources of information facilitate knowledge generation and lead to better innovation performance. Structural holes also prove to have a significant impact on innovation performance in three networks (except LCDNet). According to Burt (1992), actors in a network rich in structural holes will be able to access novel information from remote parts of the network, and exploit that information to their advantage. But in LCDNet the actors are quite close, therefore actors interact frequently. It is likely that the closeness of the LCDNet makes it less susceptible to structural holes, and thus structural holes do not influence innovation performance at LCDNet. However, structural holes do have significant impact on innovation performance at LCD-InsNet, ColNet and OptoNet. According to Burt (1992), once other actors are added to original network, the network becomes bigger in terms of the number of exchange relationships. New actors provide even more diverse information and knowledge. In such a new network, actors with structural holes have an easier time exploiting these information and knowledge to their innovation performance than actors in LCDNet.

5 Conclusion

5.1 Theoretical Contributions

Scholars have examined the relationship between network competence with innovation performance have examined the relationship between absorptive capacity with innovation performance. According to the empirical result in this research, it can be found that network competence has significant influence on absorptive capacity. Although Tsai (2001) examine absorptive capacity has moderate effect on network position, but we argue that it is worthy to analyze mediating effect between network position and absorptive capacity with innovation performance in our research model. It has evidence that network position is the mediator between network competence and absorptive capacity; network position is the mediator between network competence and innovation performance. This is full of meanings in the theoretical domain. The result will enhance the SNA role in the analysis of innovation management, including knowledge management.

The results indicate that a firm's network position indeed influences his innovation performance. Different network position variables capture different kinds of position concepts. In this study, we represent the most interpretable position variable in LCD industry, namely, coreness. It is suitable to be used in any industry which has a structure of core/periphery. Moreover, a method was proposed to observe the significant change on structural holes. The way is to add additional actors to original network and then examine the change of network indicators among each actor.

5.2 Managerial Implications

At the cluster-management-level, how to efficiently manage clusters is an important issue that government and STSP Authority must handle. Government plays as an integrity facilitator and provides related tax preferential policies in order to attract firm's investment willing. Despite STSP Authority only play an assistant role to companies, from the result of this study, the STSP Authority plays a role of communication bridge between government and companies, it offer complete "after-care service" and deeply understand every problem that companies faced in each growth stages and work them out. Accordingly, STSP Authority must make efforts in enhancing and improving relationship among LCD firms.

From the results of this study, we know that coreness has significant influences on innovation performance. Hence, LCD firms could try to communicate or collaborate with firms which are core in network. The survey reveals that the main reason for many LCD firms settled in the STSP is attaching to key firms and sucking fruitful resources from them. They can utilize the attached-companies' network position to access more information and knowledge, and enhance self-absorptive capacity and then improve their innovation performance. Therefore, at the company-level, the most important things for SMEs is how to access and stick in the network center, or at least attach to key actors.

5.3 *Limitations and Outlook*

Though the sample of LCD firms is enough (86%), return rate of other firms in optoelectronics industry is quite low (38%). It might result in statistical bias, such as the sample is less than 30. Besides, the measure of innovation performance is consists of three items, and respondents were asked to assess these items by their subjective opinions. It might also make our results bias. Moreover, to develop objective index to measure network competence and absorptive capacity in the future will be strongly commended.

In social network studies, relationships are usually specific to a particular domain. Researchers usually look at board ties, business ties, and etc. In this research, the main efforts only focus on communication relationship with collaboration, competition, and institutions. Maybe other types of relationships also have influence on innovation performance.

Future research should extend network scope. For example, future research could survey optoelectronics industry of different science park, industry, and even a country to examine effect on several subnetworks. Besides, future research should discuss actor's attributes (e.g. innovativeness) and network features at the same time and the interaction effect between them.

References

Abernathy, J.W., Utterback, M.J.: Patterns of industrial innovation. In: Tushman, L.M., Moore, L.W. (eds.) The management of innovation, 2nd edn., Ballinger/ Harper and Row, Cambridge (1988)

Ahuja, G.: Collaboration networks, structural holes, and innovation: A longitudinal study. Administrative Science Quarterly 45, 425–455 (2000)

Anderson, P., Tushman, M.L.: Technological discontinuities and dominant designs: a cyclical model of technological change. Administrative Science Quarterly 35, 604–633 (1990)

Arora, A., Gambardella, A.: Complementarity and external linkages: The strategies of large firms in biotechnology. Journal of Industrial Economics 38, 361–379 (1990)

Berg, S., Duncan, J., Friedman, P.: Joint Venture Strategies and Corporate innovation. Oelgeschlager, Cambridge (1982)

Biemans, W.G.: Managing innovation within networks. Routledge, London (1992)

Borgatti, P.S., Everett, G.M.: Models of core/periphery structures. Social Network 21, 375–395 (1999)

Burt, R.S.: Structural Holes: The Social Structure of Competition. Harvard University Press, MA (1992)

Cohen, W.M., Levinthal, D.A.: Absorptive capacity: A New Perspective on Learning and Innovation. Administrative Science Quarterly 35(1), 128–152 (1990)

Davis, F.G.: Agents without principles? The spread of the poison pill through the intercorporate network. Administrative Science Quarterly 36, 583–613 (1991)

Drejer, I., Kristensen, F.S., Laursen, K.: Cluster studies as a basis for industrial policy: The case of Denmark. Industry and Innovation 6, 171–190 (1999)

Freeman, L.C.: Centrality in Social Networks: Conceptual clarification. Social Networks 1 (1979)

Gemunden, H.G., Ritter, T.: Management technological networks: the concept of network competence. In: Gemunden, H.G., Ritter, T., Walter, A. (eds.) Relationships and networks in international markets, pp. 294–304 (1997)

Gemunden, H.G., Ritter, T., Heydebreck, P.: Network configuration and innovation success: an empirical analysis in German high-tech industries. International Journal of Research Marketing 13(5), 449–462 (1996)

Gulati, R.: Social structure and alliance formation patterns: A longitudinal analysis. Administrative Science Quarterly 40, 619–652 (1995)

Gulati, R., Garguilo, M.: Where do networks come from? American Journal of Sociology 104, 1439–1493 (1999)

Hakansson, H., Ford, D.: How should companies interact in business networks? Journal of Business Research 55, 133–139 (2002)

Hargadon, A., Sutton, I.R.: Technology brokering and innovation in a product development firm. Administrative Science Quarterly 38, 716–749 (1997)

Haunschild, R.P.: Interorganizational imitation: The impact of interlocks on corporate acquisition activity. Administrative Science Quarterly 38, 909–938 (1993)

Helfat, C.E.: Know - how and asset complementarity and dynamic capability accumulation: the case of R&D. Strategic Management Journal 18, 339–360 (1997)

Hurry, D., Miller, A.T., Bowman, E.H.: Calls on high technology: Japanese exploration of venture capital investments in United States. Strategic Management Journal 13, 85–101 (1992)

Kim, D., Kogut, B.: Technological platforms and diversification. Organization Science 7(3), 283–387 (1996)

Lane, P., Klavans, R.: Science Intelligence Capability and Innovation Performance: An Absorptive Capacity Perspective. International Journal of Technology Intelligence and Planning 1(2), 185–204 (2005)

Li, T., Calantone, R.J.: The impact of market knowledge competence on new product advantage: conceptualization and empirical examination. Journal of Marketing 62, 13–29 (1998)

Mizruchi, S.M.: Similarity of political behavior among large American corporations. American Journal of Sociology 95, 401–424 (1989)

Nooy, W.D., Mrvar, A., Batagelj, V.: Exploratory Social Network Analysis with Pajek. Cambridge University Press, New York (2005)

Porter, M.E.: Location, competition, and economic development: Local clusters in a global economy. Economic Development Quarterly 14(1), 15–34 (2000)

Powell, W.W., Koput, W.K., Smith-Doerr, L.: Interorganizational collaboration and the locus of innovation: Networks of learning in biotechnology. Administrative Science Quarterly 41, 116–145 (1996)

Ritter, T.: The Networking Company. Industrial Marketing Management 28, 467–479 (1999)

Ritter, T., Gemunden, H.G.: Network competence: Its impact on innovation success and its antecedents. Journal of Business Research 56, 745–755 (2003)

Ritter, T., Gemunden, H.G.: The impact of a company's business strategy on its technological competence, network competence and innovation success. Journal of Business Research 57, 548–556 (2004)

Ritter, T., Wilkinson, I.F., Johnston, W.: Measuring network competence: some international evidence. The Journal of Business and Industrial Marketing 17(2/3), 119–138 (2002)

Rogers, M.E., Kincaid, L.D.: Communication Networks: Toward a New Paradigm for Research. Free Press, New York (1981)

Shan, W., Walker, G., Kogut, B.: Interfirm cooperation and startup innovation in the biotechnology industry. Strategic Management Journal 15, 387–394 (1994)

Smith-Owen, J., Powell, W.W.: Knowledge networks as channels and conduits: The effects of spillovers in the Boston biotechnology community. Organization Science 15, 5–21 (2004)

Tsai, W.: Knowledge Transfer In Intraorganizational Networks: Effects of Network Position and Absorptive Capacity on Business Unit Innovation and Performance. Academy of Management Journal 44(5), 996–1004 (2001)

Tsai, W., Ghoshal, S.: Social Capital and Value Creation: The Role of Intrafirm Network. Academy of Management Journal 41(4), 464–477 (1998)

Zaheer, A., Bell, G.G.: Benefiting from network position: Firm capabilities, structural holes, and performance. Strategic Management Journal 26, 809–825 (2005)

Author Index

GPSR Compliance

The European Union's (EU) General Product Safety Regulation (GPSR) is a set of rules that requires consumer products to be safe and our obligations to ensure this.

If you have any concerns about our products, you can contact us on ProductSafety@springernature.com

In case Publisher is established outside the EU, the EU authorized representative is:

Springer Nature Customer Service Center GmbH
Europaplatz 3
69115 Heidelberg, Germany

Batch number: 10614940

Printed by Printforce, the Netherlands